W9-CNB-207

BETWEEN

XX

AND

XY

INTERSEXUALITY AND THE MYTH OF TWO SEXES

GERALD N. CALLAHAN, PH.D.

CHICAGO
REVIEW
PRESS

Library of Congress Cataloging-in-Publication Data
Callahan, Gerald N., 1946-
  Between XX and XY : intersexuality and the myth of two sexes /
Gerald N. Callahan. — 1st ed.
    p. cm.
  Includes bibliographical references and index.
  ISBN-13: 978-1-55652-785-2
  ISBN-10: 1-55652-785-3
  1. Intersexuality. 2. Gender identity disorders. 3. Gender identity. 4.
Andrology. I. Title.

  RC883.C35 2009
  616.6'94—dc22

                                                    2008040531

Interior design: Jonathan Hahn
Illustrations (unless otherwise indicated): Liv Larson Andrews

Published by Chicago Review Press, Incorporated
814 North Franklin Street
Chicago, Illinois 60610
ISBN 978-1-55652-785-2
Printed in the United States of America
5 4 3 2 1

To my wife, Gina, who makes everything possible; to my children, Jennifer and Patrick, along with my honors students—people who make me think about things like this book; and to Kailana Sidrandi Alaniz, Nicky Phillips, Lisa May Stevens, and Dianne, people who opened my eyes to the world around me.

CONTENTS

Acknowledgments

I wish to thank doctors Alice Dormurat Dreger and Joel Freder for their critical scientific contributions to and careful reading and revision of portions of this manuscript. I'm also grateful to Jane Dystel and Miriam Goderich for their encouragement and essential contributions to the development of this book. And I want to thank Cynthia Sherry and Lisa Reardon at Chicago Review Press for their interest in and support for what was clearly an unusual book proposal as well as for their contributions to the final draft.

INTRODUCTION

One morning in the fall of 1998, Lisa May Stevens, then thirty-two years old, methodically loaded her revolver. When she was done she snapped the chamber into place, pushed her strawberry blonde hair behind her ear, and snubbed the barrel of the revolver beneath her jaw. The gun metal felt cool there, reassuring. Lisa May drew a deep breath, let go of her demons, and pulled hard on the gun's trigger. The chamber rotated and brought the next shell in line with the barrel, and the hammer snapped down with all the finality of a slamming door.

And . . . nothing.

No explosion. No burning cordite. No shattered skull. No end to a life that, for the moment, just didn't seem worth living.

Lisa May was no stranger to handguns. She knew what she was doing. Trembling, she lowered the gun to her lap and opened the chamber. There it was, the deep dimple in the primer of the shell. The hammer had fallen and struck the shell just as it should have. But that was it.

A lot of things had come together to put Lisa May in this spot with this gun on this morning. Things that included rape, beatings, being shot twice, the death of her dearest love, and the ragged edge of this world. But by themselves no one of these, not even the combination together, would have left her here holding this gun. It was something much more than all of that.

Hidden inside of each of Lisa May's cells is a dark secret, a story so remarkable that it nearly shattered her life and the lives of many others, so remarkable that it still changes the lives of nearly everyone who comes to know of her secret.

Lisa May is different from most people. She has two very different sorts of cells in her body. Some look like a man's cells. Some look like a woman's cells. Because of that, some parts of Lisa May look like parts of a man, while other parts of Lisa May look like a parts of a woman.

Just a few years ago, most scientists would have referred to Lisa May as a true hermaphrodite (as Lisa May still does), but she is much more than that. She is very nearly two persons become one.

But it wasn't those odd bits of chromosomes or the mix of the male and female that drove Lisa May to her gun cabinet and her wish for oblivion. After all, those were just parts of who she was. In the end what put the bullet in her hand and the barrel to her temple was a dreadful thought filled with shame and an unimaginable sense of worthlessness, loneliness, and despair.

All of that is behind her now. Today, Lisa May embraces her differences and her hermaphroditism. But the road to her happiness was long and rough, mostly because of how many of us think about sex.

Most people quickly accept the fact that we differ in one way or another from everyone else we know. In fact, this society often encourages us to maximize, enhance, and revel in those differences. But when it comes to thinking about sex—even simple biological, boy-versus-girl sex—many of us reserve a different sort of boardroom for such thoughts. Variations among human beings that have to do with biological sex, reproduction, or genitals invoke separate sensitivities and often raw-edged discomforts.

So we try to keep sex simple. Men and women come in two, and only two, opposite forms: male and female, men and women, boys and girls. Black and white simplicity, no gray. We understand that gender—the ways that society molds us into proper girls or boys, men or women—is complicated. Gender depends on lots of things—upbringing, culture, the stories fed to us by television and movies, hormones, and power struggles. Throughout it all, though, sex remains inviolate to us—boy/ girl, black/white.

But that point of view doesn't fit very well with the world around us.

This book is about that world around us and about another way of thinking about ourselves. It began in one of my classrooms, an honors course devoted to examining how we come to have images and visions of ourselves. As I was preparing for a discussion about the importance of gender in our views of ourselves, I discovered that every year more than sixty-five thousand children are born who aren't obviously either boys or girls. I was amazed by that number. I was even more amazed by my ignorance of that number. I'm a pathologist, and I know a lot about disorders, diseases, and the things that generally make people different from one another. And I was very familiar with many disorders that affected far fewer people. How had I come so far and heard so little about these children?

That began the investigation that culminated in this book.

In the process, I have discovered some remarkable things. Our history suggests that we haven't always imagined that humans come in only two sexes, and that things far removed from what we might call facts have played major roles in determining our thoughts about sex. Even today, several human societies believe in more than two sexes.

In truth, humans come in an amazing number of forms, because human development, including human sexual development, is not an either/or proposition. Instead, between "either" and "or" there is an entire spectrum of possibilities. Some people come into this world with a vagina and testes. Others begin their lives as girls but at puberty become boys. Though we've been told that Y chromosomes make boys, there are women in this world with Y chromosomes, and there are men without Y chromosomes. Beyond that, there are people who have only a single unpaired X chromosome (people we call women who aren't exactly like other women). There are also people who are XXY, XXXY, or XXXXY whom we call men but aren't exactly like other men. There are babies born with XYY, XXX, or any of a dozen or more other known variations involving X or Y chromosomes. We humans are a diverse lot.

As I worked on the book I also met some wonderful people who were willing to share their stories with all of us—stories that too often included a litany of doctors' and families' lies and secrecy, and feelings including shame, confusion, and despair.

Because of those stories, the world around me changed forever.

I discovered that the character of our society, our language, and our past often drives us to do something about those of our children who don't fall easily into our minds as either boys or girls. The surgical and other means we have developed to help these children are amazing, but no one knows just how successful such methods are. Still, we continue to alter ourselves in sometimes painful and questionably successful ways because, I think, we believe that sexual attraction and human genitalia serve the sole purpose of human reproduction. But a single look into the animal kingdom, a lone glance at the wondrous society of bonobos, is enough to reveal that all of our preconceptions about sex are just that, preconceptions. Among other primates, sex serves a nearly unimaginable number of purposes beyond reproduction.

Because of this exploration I've come to believe that the ideas about sex that are so ingrained in us just don't fit very well with reality. Human sex is not something that switches irreversibly between two poles—male and female—like an on/off switch on a radio. Rather it is like the bass and treble knobs on that radio. Pure bass or pure treble are impossible to achieve, but in between those two exists an infinite number of possible mixtures. Inside that infinity of possibilities each one of us is nestled in the vastness between pure male and pure female.

My purpose is not to convince you that we need to imagine more sexes, because the concept of five sexes would be no closer to solving the problem than the idea of two sexes is. Instead, I wish to offer other ways of thinking about sex—ways that aren't so constraining or exclusive, ways that might even change how we think about ourselves.

1

THE PUZZLE OF INTERSEX:
THE STORY OF LENORE

It might have been one of those Los Angeles days when the soup rolled in off the sea and sopped up the sky, one of those days when people were left with nothing more than sputtering electric fans and limp palm trees curdling in the oily light. It might have been, but the report doesn't mention any of that. So it might have been otherwise.

For certain, it was the summer of 1952. Harry Truman was still president, the Dodgers were in the process of losing more road games than they've lost since, and the Ford Motor Company was preparing for its fiftieth anniversary. That summer, in the city of angels, a baby was born to two very proud parents. I will call them Frank and Laura. The baby—let's call her Lenore—was the second child in what would grow to be a family of twelve, and Frank and Laura were first cousins. Maybe that's important; maybe their genetics had something to do with the way Lenore turned out. Maybe not.

At birth, Lenore was everything everyone had hoped: all her digits present in the proper places and numbers, beautiful eyes and hair, pink gums and stubby toes. Everything about her looked perfect, with one tiny exception. Well, actually, not so tiny. Lenore's clitoris was a little too big for a baby girl. "Hypertrophied" was what the doctor called it.

But after some further probing the doctor found what seemed to be a vagina, so he announced, with a big smile, "It's a girl."

As the doctor's words split the air that day, one door opened and another one closed. No one noticed.

Wrapped in pink, Lenore went home, and for the next several years things seemed just fine. Lenore did all the things a baby girl should do. Then, when she was six years old, Lenore—like many kids her age—got the measles. Her mother took her to see the doctor. Once again, nothing about her physical exam seemed out of order, except for that clitoris thing. It still seemed a little big, but not *too* big, at least not so big that anyone felt compelled to do something about it.

By age thirteen, Lenore had begun to develop pubic and underarm hair, just as any girl her age should. Curiously, though, unlike her friends, Lenore's chest remained as flat as a boy's, and she had not begun to menstruate.

That same year, Lenore developed tonsillitis and went to the hospital to have her tonsils removed. During Lenore's intake exam, the doctor who examined her—a particularly thorough sort, apparently—again noticed the hypertrophy of Lenore's clitoris. However, unlike the others who had noted the same excess, this doctor felt the need to do something more. When he looked at her more closely, he found that Lenore's vagina was only rudimentary—too short and too small to be functional.

This, combined with Lenore's flat chest and lack of menstruation, puzzled the doctor. So he referred Lenore to the Adolescent Children's Unit of Childrens Hospital of Los Angeles for further tests. A pediatrician there performed a complete physical evaluation. In the fall of 1966, Dr. Betty Suits Tibbs wrote up the course of that evaluation, and the story of Lenore's short life to that point, in the clinical journal *Pediatrics*. Dr. Tibbs's tale is the foundation for this story. Since Dr. Tibbs's report doesn't name Lenore's physician at Childrens Hospital of Los Angeles, I'll call him Dr. Brown. Here is his description of Lenore:

> Pertinent physical findings revealed moderate hirsutism of the face [facial hair] with slight acne. The thyroid gland was not

palpable. Examination of the chest revealed a normal contour but no breast development. There were no cardiac murmurs and the heart was not enlarged. Axillary hair was ample and the pubic hair formed a female escutcheon. The clitoris was hypertrophied, appearing as a small penis with a urethral opening at its base. There was a small vaginal opening posterior to the urethral opening.

According to Dr. Brown, when Lenore appeared before him, her outward physical appearance and mannerisms seemed more boyish than girlish. In addition, Lenore's vagina was only about 2.5 centimeters deep—about an inch—and, palpate as he would, Dr. Brown could not find a uterus in Lenore's abdomen. All of this led him to suspect that in spite of her history, Lenore was in fact a boy. That seemed worth following up on, so he scraped some cells out of Lenore's mouth and sent them off for analysis.

A few days later the results of Lenore's tests came back. The genetic expectation for girls is that each cell will have two X chromosomes. Normally, early in the embryonic development of girls, in every cell of the embryo one of those two X chromosomes is inactivated and condenses into a small inert lump called a Barr body, or sex chromatin. The presence of sex chromatin usually identifies a cell as female. Dr. Brown could find no evidence of any sex chromatin in any of Lenore's cells. So, at her next visit, Dr. Brown took some blood from Lenore and sent the white cells off for analysis. The results of those tests showed that most of Lenore's cells had the right number of total chromosomes, forty-six, but the tests also showed that buried inside of every one of Lenore's cells there was an X *and* a Y chromosome. Lenore's cells were genetically male. Analysis of Lenore's hormones revealed a shortage of female hormones. It all fit, except for the fact that Lenore still thought of herself as a girl.

Because of his findings, Dr. Brown referred Lenore for psychiatric evaluation. Apparently, he wanted to find out how well Lenore's mind had adapted to life as a girl in spite of her Y chromosome.

Dr. Tibbs reported the results of those tests:

The results of the psychiatric evaluation revealed a low average intelligence. . . . The patient never actually expressed confusion on direct query as to whether she was male or female, but did demonstrate some masculine aggressive behavior patterns and fantasies. Dreams and projective tests interpreted by both the psychologist and psychiatrist also revealed some ambivalence in her gender identity. On the conscious level, the identity patterns conformed to the assigned sex.

In other words, on a conscious level Lenore acted like a girl. But subconsciously she behaved more like a boy, which fit with the way she looked, not to mention her mysteriously silent Y chromosomes.

After that analysis, it appears that Dr. Brown suggested to Lenore's parents that their daughter might be different from other children. But the parents' discomfort with the subject outweighed their curiosity. Frank and Laura seemed to have little interest in exploring the possibility of Lenore's apparent intersex condition. According to the report, Frank and Laura were "in denial." Dr. Brown's findings did, however, make Lenore's parents uneasy enough to ask if Lenore was "all boy or all girl." Dr. Brown and the doctors he worked with decided not to push the matter. They assured the parents that the physicians were just following up on the diagnosis of amenorrhea and hirsutism and that there were many possible causes of these symptoms, "rather than dwelling on the possibilities of intersex and their ramifications." Dr. Brown and his colleagues never again raised the specter of intersex with Lenore's parents.

Lenore was not the first person Dr. Brown had seen or read about with a condition like this. By the 1960s, intersex was certainly a known human syndrome, and the medical literature, albeit a patchwork of fact and opinion, contained many descriptions of intersex children. It was clear that children didn't always come in tidy packages. What was far less clear was how best to treat such babies and children. At the time, some physicians thought they could do just about anything with a child's sex. Give any baby the right clothes, the right toys, the right parents, and the right instructions, and that baby would become a boy or girl. Other physicians who studied intersex weren't so certain, especially with children as old as Lenore.

Lenore's age complicated matters. Still, Dr. Brown felt he had to act. Almost everyone accepted that something had to be done. The child had to be either a "she" or a "he." But before he made his final decision, Dr. Brown wanted to investigate further. He told Frank and Laura that the doctors needed to do a laparotomy on Lenore so they could determine why she wasn't menstruating. A laparotomy is an invasive procedure requiring a sizable incision through the abdominal muscles and some degree of poking around inside the patient's abdominal cavity. Many important things reside inside that cavity, such as the liver, pancreas, and gall bladder, so there is significant risk associated with a laparotomy. Regardless, at their doctor's suggestion, Frank and Laura agreed to the surgery. So did Lenore. They all wanted to know why she wasn't menstruating. In fact, Lenore, now fourteen, went so far as to thank the doctors for "making themselves available to her at this time."

During the exploratory surgery, the doctors discovered that Lenore had two undescended testes. They also found a uterus of sorts and rudimentary Fallopian tubes. That confirmed everything Dr. Brown and the others suspected—that Lenore, who had been raised for fourteen years as a girl and whose genitalia were mostly those of a girl's, in fact had not only the chromosomes but also the gonads of a boy.

Some minds were probably made up even before the surgery. But now the doctors agreed on the spot: Lenore's life might just as well continue as a girl's life. So during the laparotomy, the doctors took it upon themselves to remove Lenore's testes. They later told Lenore and her parents that her gonads were abnormal, so the doctors removed them. It isn't clear whether the doctors ever told the parents or anyone else that those "abnormal" gonads were, in fact, testes and not ovaries. They did say that it was because of this abnormality that Lenore did not menstruate. Once again the doctors assured the parents that there was no doubt about Lenore's sex. The orchidectomy, or removal of the testicles, was the first step in shoring up that reassurance. Further procedures would set it in concrete.

Three months after the laparotomy, the doctors took Lenore back into surgery (the report is a little vague about the justification for this surgery) and performed a clitorectomy; that is, they cut off Lenore's too-small penis or too-large clitoris (whichever it was) to make her look more feminine. Then Dr. Brown administered a full course of estrogen therapy.

To the doctors, it seemed that the only remaining problem was Lenore's lack of ovaries. Simply by providing her with the hormones that nature had denied her, the doctors could give Lenore that final push down the road to womanhood. In part, at least, the doctors were right. Injections would provide Lenore with the estrogen she lacked and stimulate secondary sexual development, especially breast formation. But beyond that no one knew for sure what all those hormones might do.

It appears that each of the physicians involved felt certain that Lenore's best future was as an assigned and surgically simplified, chemically enhanced female, regardless of her Y chromosomes and her testes. We can't know all of the factors involved in their decision, but it's probable that Lenore's fourteen years spent as a girl played a part. It's also likely that the difficulty of reconstructive surgery to make genitalia appear more masculine influenced them as well. And perhaps there were other factors. One thing is clear in Dr. Tibbs's report: no one considered the option of leaving Lenore as they found her. Lenore, as she came to them, didn't fit into either of the two spaces human minds reserve for human beings—she wasn't clearly a boy or a girl. Empathy, morality, and kindness surely played a role in the doctors' decisions, but so too did a certain deep predisposition.

Even though Lenore had been swimming in a pond of male hormones for over a dozen years, the doctors scalpeled and hormoned Lenore—a 46,XY person—into a girl. With the hormones given her, Lenore's facial hair cleared up and her breasts began to grow, gracing her chest with one of the most noticeable badges of womanhood.[1]

But of course none of this really solved the problem, if in fact there ever was a problem that needed solving. In the end, to some Lenore was clearly a girl. At the same time, to others Lenore remained a boy. But Lenore was neither.

What, then, should we call Lenore?

Dr. Betty Suits Tibbs's report ends with Lenore in the tenth grade—maybe sixteen years old. Dr. Tibbs observed, "The patient is in tenth grade at present and has made a very good adjustment. It is felt that with her drive and capacities, the prognosis for her identity as a woman is quite good." The rest of Lenore's story we can only imagine.

Every year in the United States, approximately one thousand babies are born with cystic fibrosis and about four hundred are born with hemophilia. Few of us have to ask what hemophilia or cystic fibrosis are. We might not fully understand what causes these disorders, but we know that either one can make a person's life very difficult. Curiously, no one seems to know just how many babies of indeterminate sex like Lenore are born in the United States every year. Estimates range from one thousand to fifteen thousand. It seems probable that the correct number is nearer to the lower estimate than the higher one. Regardless, it is a substantial number of people. We now refer to these people collectively as intersex, or people with disorders of sex development (DSDs). The birth of an intersex child is a difficult event for family and physicians. They must select, from very few options, the least-bad alternative with the hope that, even in their ignorance, even with the paucity of language available to speak about these children, even under the weight of history and fear, they may create a better future for their new child.

Surprisingly, until very recently, standard practice usually excluded the child and the parents from the decision-making process. The physicians made the choice of boy or girl and did what they could do to ensure that the child would walk that path for the rest of his or her life. Physicians believed they knew best and that the input of others was unnecessary. They believed that the fewer who knew about what had happened at birth, the less likely it was that someone might raise the veil so carefully woven and placed by their hands.

As a result, for years hardly anyone outside the medical community had heard of the thousands upon thousands of children like Lenore.

And even today, most of us hear little about these people who, like Lenore, fall through the cracks in our language and raise serious questions about our cast-iron ideas about two opposite sexes. Maybe we don't hear much about these people simply because we don't *want* to hear about them. They make us even more uneasy about things we are already sufficiently uneasy about—things like human sex. But regardless of our discomfort, these are people, and their stories are important, because wrapped up inside of them is a secret that all of us should know, a secret about what it truly means to be human.

2

A BRIEF HISTORY OF SEX

In many ways Lenore's future was laid out years before she was born. As Dr. Brown and his colleagues pondered Lenore's situation, millennia of human thought about sex and intersex molded their ideas. As long as human beings have walked upright, we've been thinking about how we acquire sex, how we have sex, and why we need sex. And though most of us may believe that humans have always thought in more or less the same ways about the sexual character of human beings, history doesn't support that assumption. In fact, we have not even always believed that humans come in two, and only two, opposite sexes.

Ancient Greek Sex (c. 450 B.C.–A.D. 200): The Power of One

Our access to musings about the biology of human sex begins about twenty-four hundred years ago with people like Hippocrates (c. 460–c. 377 B.C.), Aristotle (384–322 B.C.), and Galen (A.D. 129–c. 199). Hippocrates, the father of medicine, forced the first separation of religion and science. He argued for the natural origins of disease and death and shunned the gods. He was a logical man and a careful observer of humankind. Nothing he proposed originated from fancy or spur-of-the-moment decisions. He was thorough and methodical.

9

Hippocrates proposed that menstrual blood and sperm were in essence the same substance. Women shed menstrual blood, he said, when an excess of nutrients accumulated in the blood. Men, instead, refined blood foam into sperm and passed it along to the brain. The sperm then made its way through the spinal marrow, into the kidneys, to the testicles, and finally into the penis itself. Hippocrates clearly saw differences between men and women, but for him the differences were of process—like subject and exposition, parts of a single fugue—not basic differences in nature, and certainly not opposite sexes. He even proposed that both men and women produced sperm, or seed, and that only the strength of their seed differed. Even then, he said, sometimes men produced strong sperm and at other times weak sperm, and the same was true for women. Men, of course, because they were stronger, came from the strong sperm. Nevertheless, to Hippocrates, men and women came from a single mold.

Aristotle, whose life overlapped briefly with that of Hippocrates, studied under Plato and schooled Alexander the Great. Aristotle wrote about nearly everything: philosophy, physics, metaphysics, poetry, theater, music, logic, rhetoric, politics, government, ethics, biology, zoology, and sex.

Aristotle's notions about sex appear to be divided. At times, his basic philosophy seemed to commit him to the two-distinct-sexes model of men and women. But his writings often expressed a different view. Aristotle certainly didn't think that men and women were identical, but neither did they appear in his writings as polar opposites. The major differences between men and women, he said, involved the internal and external genitalia (both were parts of the gastrointestinal system in his anatomy) and the roles men and women played in procreation. He wrote, "The female always provides the material, the male that which fashions it, for this is the power we say each possess, and this is what for them is to be male and female. . . . While the body is from the female, it is the soul that is from the male."[1]

Likewise, according to Aristotle, a castrated male became essentially a female. The womb was the distinctive part of the female, and the penis the distinguishing male feature. Otherwise, they were mostly alike. The differences in anatomy allowed for differences in function, but not differences in kind. Aristotle even went so far as to suggest that the womb was in essence the homologue of the scrotum. Men and women were not two poles of humanity, but one people provided with essential, minor differences that allowed for procreation.

Five hundred years later, in the second and third centuries, Galen, in the Greek city of Pergamum, was the single most influential physician and anatomist in the world. He changed nearly everything everyone thought about human anatomy and medicine. Many of his thoughts about these disciplines would dominate human medicine for the next fifteen hundred years, some for even longer.

Medically, Galen was a true radical. He successfully performed bold surgeries that no one else would dare to duplicate for nearly two millennia. He performed cataract surgeries by inserting needles into his patients' eyes. He routinely opened up people's skulls and worked on their brains. Galen revolutionized nearly every aspect of medicine that he touched—with one notable exception.

Galen wrote at length about the differences between men and women, but, unlike most of his other endeavors, in this arena he came to think much like those who had come before him, and his conclusions were more or less the same as those of Hippocrates and Aristotle—that humans truly are not so very different from one another. Galen proposed, in essence, that women were simply men turned inside of themselves:

> Think first, please, of the man's "[external genitalia]" turned in and extending inward between the rectum and the bladder. If this should happen, the scrotum would necessarily take the place of the uterus with the testes lying outside, next to it on either side.
>
> Think too please, of . . . the uterus turned outward and projecting. Would not the testes "[ovaries]" necessarily then be inside of it? Would it not contain them like a scrotum? Would not the neck "[the cervix and vagina]," hitherto concealed inside of the perineum but now pendant, be made into the

male member? . . . you could not find a single part left over that had not simply changed its position.[2]

In other words, we were first of all human beings, with all the same parts. Our only differences, according to Galen, arose from variations in the arrangement of those parts.

The Greeks' ideas about sex may seem strange or silly to us today, but the accuracy of their beliefs is not the point. It is how the ways we have thought about sex have changed over time that is important. Absurd or not, the Greeks' views remained dominant among physicians, scientists, and laypeople alike until around the time of the Enlightenment in the eighteenth century.

Renaissance Sex: Leonardo da Vinci (1452–1519) and Andreas Vesalius (1514–1564)

One way to examine how our thoughts about sex have changed is to compare anatomical depictions from centuries past with those of modern scientists. Some of our greatest windows into our past are the drawings and paintings of Leonardo da Vinci. Leonardo was a curious man, in almost every sense of the word. It was curiosity that drove him to split open human corpses and lift the corrugated tubes and gristle and muscles and bones and organs from within those cadavers and line those pieces up neatly along the edges of his oiled dissection table. Curiosity, too, drew him into thoughts of the sky and flying machines, tanks, submarines, the Mona Lisa, and the Last Supper. And curiosity finally compelled him to seek the underpinnings of human sexuality and the anatomy of sexual intercourse.

For years, Leonardo da Vinci had studied human anatomy—patiently slicing through skin and muscle, bone and viscera, cracking skulls and puncturing eyeballs—undressing the dead in the most intimate of ways. On top of that, he was a meticulous observer. Nothing escaped his attention. By the low lights of his candles he drew each of his dissections and made careful notes in the pages of his notebooks—every tendon, every fascia, each nerve found its way there. In his work he was methodical, meticulous, nearly maniacal in his search for the truth. He knew human bodies like few others. Leonardo's anatomical drawings rank among the most beautiful ever made.

Leonardo's lifetime of work led him to envision a great treatise on the nature of human beings—a book that would place him alongside Galen as one of the greatest anatomists of all time, a book that would lay the lives of men and women open like tapestries colored only by the fluids that make us human. This book would begin with the intricacies of human sex and the moment of human conception; after all, this was where each life began. Then he would work his way through the complexities of life in the womb, a human birth, a child, a man, a woman, each with all the nerves and muscles and bones and blood vessels. The book would end, as each of us does, in the contradictions of old age and the wrinkled face of death. It would be the culmination of his life's work.[3]

Leonardo da Vinci would have his great book, and it would open with something no one had ever seen or even attempted before.

Inside his studio, Leonardo began to sketch. A man he had known, perhaps, or maybe all the men he had known took shape beneath his fingers as he worked across the page. But not a whole man. Instead Leonardo laid out for our inspection a man cut, from skull tip to scrotum, in half. Once he had the man's outline in place, Leonardo's hand pulled out the shape of a heavy-breasted woman, headless and split like a gourd from pubis to clavicle.

Spines and hearts, bones and brains fell onto the page. Muscles took shape and backs stiffened. The last part must have presented a particular challenge. Penises he could draw, he'd seen dozens, even in cross-section, and vaginas as well. But he had never seen this moment as an anatomist, from the inside. He was curious, though. He added the erect penis, opened the vagina slightly, and pulled the two together.

"I expose to men the origin of their first, and perhaps second, reason for existing."[4] With those tantalizing words Leonardo unveiled one of his now lesser-known drawings: *The Copulation*, a large, elaborate sketch of a man and a woman engaged in the most essential union of humankind—sexual intercourse.

Though Leonardo surely wasn't the first to wonder what sex might look like, he was the first, so far as I know, to draw in such detail two humans so intimately coupled. Curiously, the part he knew least about is the part he got most nearly right, or so it seems from our present perspective. The union of the man and the woman—the part he wrought from his imagination—lies mostly correct upon his page. The man's penis falls nearly where it should within the woman's vagina. The shape and placement are off a little, but not by much. The depth of penetration, the placement of the organs, and the angles of the bodies all are nearly correct. But the part Leonardo knew, or should have known most fully—the simple truths of human anatomy—is rich with errors.

From the woman's uterus, a duct extends into her abdomen, passes through her chest, and ends inside the nipple of a single breast. To modern-day anatomists it seems no such duct exists. In the man, another duct extends from the penis and reaches, via the spinal cord, to the brain. People have speculated that Leonardo put it there to carry sperm from its birthplace, which he believed was inside a man's brain, and deliver it at just the right moment to the tip of his penis. The man's penis contains two tubes—one apparently for urine, the other for sperm.

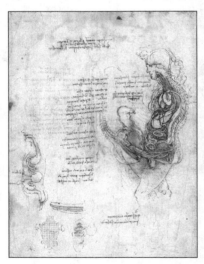

Of course, many have proposed a direct connection between a man's penis and his brain, but neither of these ducts appears in modern anatomical drawings. Sperm comes from the testicles, not the brain, and a single tube carries both sperm and urine, though not simultaneously, through the shaft of the penis.

Finally, in Leonardo's rendering there is another tube, a blood vessel perhaps, that runs directly from the penis and scrotum to the heart. Possibly he put it there to provide the blood needed for the man's erection. Perhaps he put it there to tie sex to love. Regardless, no one today sees that tube.

The Copulation: a sketch by Leonardo da Vinci circa 1493. *Reproduced with permission from the Royal Collection, London, England*

Like Galen before him, Leonardo saw things that we no longer see, things we now believe never existed. How could that have happened? How could an anatomist as skilled as Leonardo, a man so painstakingly accurate at other times, have made such gross errors? Of course, the short answer is that no one knows, and no one ever will. Whatever motivated Leonardo died with him.

But it is interesting that, at the time, Catholic Church doctrine fit a lot better with Leonardo's image of two tubes through the penis, which would keep pure sperm "unpolluted" by urine. In addition, over a thousand years before Leonardo, Hippocrates had proposed that the brain produced sperm and delivered it to the penis, even though Hippocrates never saw such a duct either. The Church also liked that idea, because it separated the holy act of reproduction from the base function of elimination. Old ideas die hard, and political ideas backed by the strong arm of governing religious institutions fitted with the talons of the Spanish Inquisition also acquire a certain attractiveness they might otherwise lack.

One way or another, Leonardo clearly drew something that those who look today cannot see.

Only a few years after Leonardo's death, another young man drawn by the siren song of anatomy set upon the task of revolutionizing the science and once and for all laying to rest the ignorance of the ancients. That man was Andreas Vesalius, who, at the peak of his career as anatomist and physician at the University of Padua in Italy, embarked on an assault on the work of Galen. To Vesalius's great good fortune, a local judge took an interest in the anatomist's work and saw to it that the bodies of executed criminals found their way into Vesalius's dissection theater. Vesalius hired an artist, probably Jan Stephan van Calcar, a student of Titian in Venice, to draw from the dissections. Because of this artistic input, the resulting drawings were superior—more accurate and intricate—to any that had come before. In 1555, Vesalius gathered van Calcar's drawings into the classic work of anatomy *De Humani Corporis Fabrica*.

The *Fabrica* represented the first major change since Galen. In fact, many of Vesalius's findings directly contradicted those of Galen. So great were those contradictions that some of the most learned scholars of the time, who revered Galen, severely criticized Vesalius for the *Fabrica*. Vesalius replied that the ones who deserved the criticism were those who mindlessly followed the works of Galen and couldn't see Galen's errors for themselves. Despite the critics, the *Fabrica* quickly became regarded as the first truly accurate representation of human anatomy. Finally, people thought, the anatomists had it right. The ignorance of the past now lay buried beneath the rich soil of modern truth.

But among all of Vesalius's criticisms of Galen's work lay a single nugget of agreement—the similarity of human genitalia. Vesalius and van Calcar laid out the vagina as a long tube that looks very much like the shaft of a penis. At the external end of the vulva the tissues are gathered into the shape of the glans of the penis and covered with pubic hair. At the opposite end, the uterus acquires the shape of a scrotum.

The similarity of the sexes in these drawings was apparent to all, and when another anatomist, Juan Valverde de Amusco, published his *Historia de la composición del cuerpo humano* in 1556 in Venice, he followed Vesalius's lead. In his drawings, women's genitalia also looked remarkably like those of men.[5]

From Vesalius and his disciples this vision spread. Nearly every drawing of women's genitalia for the next three hundred years would look like those first created by Vesalius and Valverde de Amusco. For most of the first two millennia of recorded history, men and women appeared more alike than they ever have since.

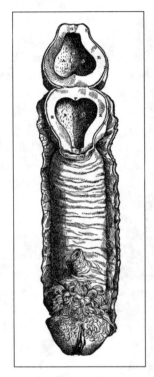

The female genitalia as seen by Vesalius, circa 1540. The uterus is at the top, the vagina and vulva below. *The Illustrations from the Works of Andreas Vesalius*, World Publishing Company.

Believing Is Seeing

Looking at some of the drawings of Leonardo, Vesalius, and Valverde de Amusco, my first temptation is to believe that the anatomists who created them simply lied about what they saw—just deleted, reshaped, and created things to suit their expectations and popular beliefs. But these were some of the best-trained and most critical observers alive. They sought the truth, especially Leonardo. I cannot easily believe that he would have intentionally lied about what he saw simply to pay homage to currently popular ideas among anatomists and the whims of the Catholic Church. Nor can I believe anything similar about Vesalius, who seemed often to delight in displeasing others. That leaves me with only one option: I have to believe that these drawings reflect what these men truly saw—men and women as very nearly alike.

We cannot peer into the minds of men and women dead for nearly seven hundred years. But we can learn from them. The amount we can learn, though, depends on our ability to set aside our own prejudices and the things we now believe to be true. Seeing does not happen simply. We don't just open our eyes and "see" whatever is out there. Our minds contribute greatly to the process.[6] The act of seeing involves mental processing—our brains filter, tweak, color, and manipulate our visions, a little like what Photoshop can do to the "reality" of a photograph. Things get added and deleted, colors change, backgrounds sharpen or dim, perspectives change.

Center for the Study of Intelligence, Central Intelligence Agency, 1999.[7]

The image here seems simple enough—a few triangles with a few familiar phrases stacked inside those triangles. But it isn't simple. At first glance, almost no one notices that the articles are repeated in every one of the triangles. Only after someone points it out do we see them. The combination of the familiarity of the sayings with the repeating triangular shape of the presentation leads us quickly through the words. We don't intentionally leave out the extra words. We don't feel any religious or political pressure to ignore the extra words. We actually don't see them, and we don't see them because we don't expect to see them.

Similarly, as long as people believed and expected that men and women were more alike than different, as long as it seemed sperm should come from the brain, then that was what people uncovered when they poked around inside the dead. The ideas were the important thing, the corpses and dissections were needed only to confirm those ideas. From Hippocrates to Vesalius, for over two thousand years, that's how things stood with human sex, until the Western world found itself on the verge of the Enlightenment.

Columbus Discovers the Clitoris: The New World of Sex

So how did things change? How did the two-sex model of human beings displace the old one-sex way of looking at things? Many factors probably played a role, but it seems likely that the switch in viewpoint didn't result from any major change in human nature or groundbreaking scientific discovery. It also seems likely that one small event in particular played a surprisingly large role.

In 1546, a young anatomist working at the University of Padua, Matteo Renaldus Columbus (or Realdo Colombo), who had worked for years with Vesalius, was suddenly elevated to director of the Institute for Anatomy. His career was one of considerable brilliance. Among other things, Columbus showed that the blood flowed from the right side of the heart into the lungs, that the lens of the eye lay at the front of the eyeball, not in the middle (as many believed), and that the arteries expand with each contraction of the ventricles.

Successful, handsome, heavily bearded, and hardworking, Columbus had come far. But in his mind none of his accomplishments equaled the one he was about to announce. Near the end of his life, Columbus published his own work of anatomy, *De Re Anatomica*. And after fourteen years of study at the University of Padua, with a drum roll and trumpets, Matteo Renaldus Columbus announced that he had discovered the clitoris.

Breaking two traditions at once, Columbus had used a living person (we can only assume to his wife's delight) for his studies, and he had relied on his own observations rather than theory and historical precedent.

"It is 'preeminently the seat of a woman's delight.' Like a penis, 'if you touch it you will find it rendered a little harder and oblong to such a degree that it shows itself as a sort of male member. . . . Since no one has discerned these projections' workings, if it is permissible to give names to things discovered by me, it should be called the love or sweetness of Venus.'"[8]

Columbus was immensely proud of his "discovery," but it immediately created controversy. First, and most important, it seemed to contradict the one-sex hypothesis then popular. Now, according to Columbus, a woman had an exterior counterpart to a man's penis. How could a woman have "two penises" and still be the perfect homologue of and basically the same as a man? That rattled the foundations of then-current thought. But once everyone took a careful look, they had to agree—the clitoris was in fact there, and it did seem a lot like a little penis. That began to work at people's minds. Where certainty had ruled for nearly two thousand years, a seed of doubt began to sprout.

Surely there were other seeds sown along the way, but in spite of its small size the clitoris was one of the largest stones hurled against the seemingly impenetrable wall of the one-sex worldview. Slowly, we began to think about ourselves differently.

The second controversy Columbus's announcement created was a furor among his colleagues, who attacked him viciously—not because of the absurdity of his claim, but because *they* wished to take credit for the clitoris. Several counterclaims were immediately issued. The most notable among these came from Columbus's colleague, Gabriel Falloppius—who discovered the Fallopian tube. As soon as Columbus retired and Falloppius succeeded him at the University of Padua, Falloppius claimed that he, not Columbus, had discovered the clitoris and that Columbus and everyone else were plagiarists.[9]

Many of Columbus's discoveries described in De Re Anatomica overlapped the discoveries of Falloppius. Falloppius had published his own book of anatomy, Observationes Anatomicae, in 1561—shortly after Columbus's death—and claimed he had completed his book four years before Columbus's, which is probably untrue. Nevertheless, over ten years after Columbus's death, one of Falloppius's students, G. B. Carcano, formally charged Columbus with plagiarism. (Columbus, of course, ignored the charges.)

The controversy over who deserved credit for discovering the clitoris raged on until one hundred years later, when a widely known and respected anatomist, Kaspar Bartholin—a professor of medicine and later of theology at the University of Copenhagen and author of his own textbook of anatomy, Institutiones Anatomicae—said that both Columbus and Falloppius were foolish for having claimed the "'invention or first Observation of this Part,' since the clitoris had been known to everyone since the second century."[10]

Four hundred years after that, following a discussion of these issues in one of my classes, several of my female students stated very convincingly that they think it probable that the clitoris was discovered long before the second century, and most likely it was not a man who first found it. Furthermore, they've told me that this tale makes some of them feel as Native Americans must have felt when the other Columbus declared he had discovered a "new" world.

All of this would be nothing more than a fascinating look at anatomical history and male hubris except for one thing. After a brief stint of popularity through the second and third centuries, the clitoris had, in fact, fallen off the map. After Galen, most anatomical drawings lacked clitorises. And for the most part the clitoris stayed off the map until Columbus, Falloppius, and others decided the clitoris was there and wrote their "discovery" down. Amazingly, clitorises then began appearing in all sorts of anatomical drawings. Apparently those who were dissecting women's bodies and making drawings of them couldn't see a clitoris until they were told that it was there.

But while the clitoris had reappeared on the scene, and the direction of human sexuality had changed forever, in the short term, ideas about much of the rest of women's anatomies remained mired in the bog of popular opinion. Da Vinci's drawings—with the vagina and the uterus still in the shape of a penis, and the vas deferens (the ducts that transport sperm from the testicles to the penis) looking surprisingly like uteruses—still held sway, and so did the drawings of Galen created hundreds of years before. The seeds Columbus planted were sprouting, but popular opinion held strong. Only time and the deaths of many old anatomists would lead to true change in our views of the sexes.

Eventually, by the beginning of the eighteenth century, the one-sex theory finally began to crumble. Along with that change came changes in our language. Before the eighteenth century, men and women were thought to be so much alike physically that no one thought we needed different words for our similar parts. Now words were invented for the vagina and the uterus, previously unnamed because of their obvious homology to the male penis and the scrotum. The same was true for the ovaries, known until then as female testes.[11]

It is impossible to name with certainty all of the forces behind the shift from seeing ourselves as one to seeing ourselves as two. Some people have argued that the roots are political, others that it was a misguided effort to somehow link physiological differences to social differences—to tie sex to gender. But what we can know is that once human beings thought of themselves very differently than we do now, and that once we were a little more taken with our similarities than with our differences.

Sex and Sexual Intercourse: A History of Obsession

As our ideas about the biology of human sex have evolved, so have our ideas about sexual intercourse. After all, our fascination with our genitalia stems, in part at least, from our obsession with sexual intercourse. Four hundred and forty years after Leonardo da Vinci, R. L. Dickinson made another attempt at depicting the intricacies of human copulation.[12] To create his canvas, Dickinson inserted a large glass tube, as big around and as long as an erect penis, into the vaginas of sexually aroused women, and from ten firsthand insights he drew what he imagined would be the position of the genitals during a natural act of sexual intercourse. There are no obvious errors, no serious deviation from what we now consider to be true human anatomy. Shown in cross-section, the penis extends deeply and very nearly straight into the vagina. That, it turns out, is wrong. But working with glass tubes and only half the partnership, Dickinson could hardly have done better. That was nowhere near the end of this investigation.

Thirty or so years after Dickinson peered through his tubes, William Masters and Virginia Johnson took another shot at it. They used a mechanical penis that could artificially simulate coitus and direct observation using a speculum and "bimanual palpation," or manual examination of the internal organs from both inside and outside the body.[13] Their most notable findings included the observation that, during intercourse, the uterus enlarged and the vaginal walls shifted, things not seen by any earlier observer.

Unconvinced by Masters and Johnson's findings, British physician A. J. Riley and his wife used ultrasound technology in 1992 to further investigate the anatomy of coitus.[14] Unfortunately, the pictures they produced were of very poor quality—largely because they used inexpensive equipment and carried out all their scans on themselves while having intercourse—and their studies provided little in the way of further useful information.

The most recent of these sorts of studies took place in 1999, in Groningen, the Netherlands, when gynecologist Willibrord Weijmar Schulz, physiologist Pek van Andel, anthropologist Ida Sabelis, and radiologist Eduard Mooyaart used magnetic resonance imaging (MRI) to attempt to definitively and finally reveal the secrets of human mating. They found that "during intercourse in the 'missionary position' the penis has the shape of a boomerang, and 1/3 of its length consists of the root of the penis. During female sexual arousal with intercourse the uterus was raised and the anterior vaginal wall lengthened. The size of the uterus did not change during intercourse . . ." even though Masters and Johnson swore it did. Schultz and his colleagues don't say how they got those couples into an MRI tube, how they managed to get any clear images at all, or whether all of that may have affected the relative locations of any the important parts. But for the moment it seems to be the definitive study.[15] However, we've already seen the sorts of problems that arise when something gets labeled "the definitive study."

Interestingly, in none of these studies did the authors think to question whether their findings and conclusions would be meaningful for more than the one man and one woman who had participated in the particular study. No one ever asked whether the obvious differences in genitalia among men and among women might have some relevance, or considered that the anatomy of intercourse might be as varied as the men and women engaging in it. Just as we have come to think of biological sex as offering only two opposite options, we have come to imagine that sex between any man and woman is just like sex between every other man and woman.

You'd think that after all these years, after all the drawings and pictures, the ultrasounds and MRIs, we would have it right by now. Yet the sexual anatomy of human beings remains an evolving and sometimes contentious concept.

Changing Times, Changing Sexes: Science as a Moveable Feast

Or rather, sexual anatomy and the differences between men and women *were* evolving concepts, right? Now we have it correct, and the story has stopped evolving. Today we have a place for everything, and everything is in its place. The curlicue of the clitoris nestles in its proper crown. The ball bearing of the testicle lies oiled and sheathed beneath the caliper of the epididymis. All is right and fixed in the anatomical world, isn't it?

Perhaps not. For example, nested just above human and most other mammals' hearts is an organ called the thymus. In immunology there is no organ as singularly important as the thymus; within its membranous walls, our immune systems learn the mystery of self/non-self discrimination. Without thymuses, humans and other animals disintegrate under the onslaught of infection. If you transplant skin from, say, a normal chicken onto a normal mouse, within just a few days the mouse will mount a violent immunological attack on the chicken skin and reject it. If you perform the same experiment using normal chicken skin and a mouse without a thymus, the mouse will grow feathers. Mice without thymuses have lost the ability to distinguish themselves from chickens. That is a very serious sort of identity crisis.

Obviously something nearly miraculous happens inside of mammalian thymuses. This organ's role is so crucial to the developing immune system that for the past thirty years immunologists have studied no other organ so intensely. Much of what we now claim to know about how the immune system works we've learned from thymectomized mice (mice from which the thymus was surgically removed).

In 2006, though, a funny thing happened. Hans-Reimer Rotewald—working at the University of Ulm in Germany—discovered that mice have more than one thymus. In addition to the thymus near the heart everyone knew about, Rotewald and his coworkers discovered that mice have an additional thymus or thymuses in their necks—extra thymuses that everyone had overlooked all this time.[16] Well, not all this time—in the 1960s, Lloyd Law, working at the National Institutes of Health, had shown that mice had other thymic material in their necks. But almost as soon as Law published his findings they were forgotten, mostly because they were an inconvenience. As a result, I worked for over thirty years in immunology and never heard a word about extra thymuses.

Then Rotewald proved that neck thymuses work every bit as well as near-heart thymuses, and that changed the way everybody thought about most everything immunological—as recently as 2006. That fractured my once-fervent belief that anatomy was a dead science and that we had little to learn from any further study of the appearance, shape, contours, and crevices of the human body. Similarly, less than five years ago, any knowledgeable neuroscientist would have told you unequivocally that brains don't make new neurons. You are born with all you will ever have, and that's it. This is absolutely wrong.[17] And until just a few months ago, we were told that most mammals began life with all of the eggs they would ever have, and for at least some mammals that is equally wrong.[18]

Regardless of these discoveries, most of us—especially those of us in science—believe that we've finally got it right. We look at research done twenty or certainly one hundred years ago as we might look at a confused ancestor—affectionately but without belief. The efforts of last century's scientists appear almost humorous; they meant well, they just didn't know any better. Never do we imagine that fifty or one hundred years from now scientists may view with equal amusement the work we do today—the foolishness of our theories and hypotheses about the world and the human body.

I no longer assume that our perception of the sexual anatomies of males and females is static and beyond revision. If such important things as an extra thymus or thymuses or new brain neurons could be ignored into the twenty-first century, what else have we overlooked? What more will we still find among the human rubble we have been staring at so persistently for so long?

Another Side of Sex: A Brief History of Intersex

While Hippocrates and Galen and Vesalius and Columbus were offering their opinions about what people really looked like, another story ran like old water underneath these anatomists' tales—a story of other sexes and other lives, a story that still threatens to topple the house of cards we have built on the unsteady table of the two-sex life.

While most of us find talk of sex a little unsettling, we do speak of it. The meaning and the determination of the sexes, along with the importance and inevitability of intersex, though, we tend to bury in the back pages of medical textbooks.

It has not always been so. The Old Testament tells of how men and women arose from a single being. Perhaps Adam himself was the first androgyne. And as long ago as two or three hundred years B.C.E., Greek and Roman tales told of children born with parts of both sexes.[19] Often these children were thought to be oracles, and their births—surrounded by phantoms—portended catastrophic events for the empire.

Greek mythology also tells the story of Hermaphroditus, the androgynous offspring of Aphrodite and Hermes, who would lend his name to generations of intersex people to come.

From the moment he laid eyes on her, Hermes was smitten with Aphrodite. Soon after they wed, she gave birth to many sons. One of them, Hermaphroditus, was a strikingly handsome boy. For the first fifteen years of his life, Hermaphroditus lived with the Naiads in the caves of Mount Ida on Crete. He grew restless, though, and set off to discover unknown lands. Near the land of Lycia, he came upon a pool of water so crystalline he could see to the bottom. He stood and stared, for he had never seen water such as this. Salmacis, a nymph of the pool, saw him there and was so stricken by his beauty that she pleaded with Hermaphroditus to marry her. He refused and threatened to leave if she didn't stop. Salmacis withdrew, uttering apologies, and hid among the bushes.

Hermaphroditus, feeling the day's heat, stripped to his skin and waded into the water. Salmacis watched from her hiding place, and once Hermaphroditus was fully in the water, she too stripped off her clothes and dove in after him. Before he knew what was happening, she was on him, wrapping her arms and legs around him and clinging to him tightly. He tried to get away, but she was strong. As Salmacis clung to Hermaphroditus, she prayed to the gods that the two of them would never be parted. The gods heard, and in answer to her prayer, Salmacis and Hermaphroditus became one being—half man and half woman—the first human hermaphrodite.[20]

In spite of their godly beginnings, among common folk the concepts of androgyny and hermaphroditism remained anathema. In ancient Greece, the birth of an androgyne threatened the social order. The birth might be a warning that all of humankind was about to change or, at the very least, deliver into this world a creature with no clear social position. Either way, such a child opened a crack in the civilized world. Men and women had very clearly defined social, sexual, and political roles. Androgynes did not. So the good citizens of Greece created rules about what do when a child of unusual sex fell into their lives.

First, something had to be done with the child; it was an offense to the gods, not to mention friends and family. Usually the child was drowned, preferably in the sea. As soon as possible after the birth of an afflicted child, people who understood what must be done lifted the living child into a crude basket, carried the baby out to sea in a boat, and cast the child overboard.

Once again ashore, reparations had to be made. According to one oracle these should include:

1. a collection of money to offer to Demeter (the goddess of the harvest)
2. the sacrifice of twenty-seven bulls
3. the sacrifice of white cows by twenty-seven girls and prayers said by the same twenty-seven girls, according to a Greek rite in honor of Hera Basilissa (Juno the Queen)
4. an offering made by maidens (a daily libation)
5. an offering of torches for Demeter
6. another offering by matrons, with a triple libation for Demeter
7. a similar offering to Persephone (the Queen of the Under world)
8. another collection of money for an offering

And that was the minimum needed for a people's salvation. Clearly the birth of an androgyne was on par with the appearance of a comet and not something these people or their gods could take lightly.

Things remained pretty much the same until the time of the Roman scholar Pliny the Elder (A.D. 23–79), over two hundred years later. Pliny was among the first to use the word *hermaphrodite*, after Hermaphroditus, for beings who seemed to fall between men and women. Pliny was also among the first to suggest that these people were just that, people, who suffered from an accident of nature—not oracles or ill omens.

During the Enlightenment—nearly sixteen hundred years after Pliny the Elder had done his best to lift up the light of reason and scatter the darkness of myths and preconceptions—things shifted. Oracles and ghosts had fallen into disfavor. Ambiguity and indeterminacy were no longer options. Children born into that enlightened world had to be either boys or girls, regardless of how little they might resemble either. Now, rather than cast an ambiguous child out to sea, doctors, midwives, fathers, and mothers simply took a look between a newborn child's legs and chose a sex. Sometimes that required a moment or two of reflection, or a stretch of the imagination, but it was never impossible. Everyone needed and received one of two sexes.

Sex Under the Microscope

In the 1800s, histologists (those who first used a microscope to probe the depths of human tissues) and anatomists raised their lamps and offered to show us another path through the labyrinth. Eyes, which had served us so well for so long, just weren't enough. As Descartes had shown, our senses could deceive us. The faint candles of our eyes could light only the shallowest of this world's dark pools. Now people and their parts could be dissected and deconstructed under the unflinching eyes of the microscope, opening up an entire universe beyond the curtains of our eyesight. Nothing was exempt from these scientists and their tools, not human anatomy or physiology, not reproduction, certainly not sex, not even the sticky problem of sex assignment. The microscope told all. Testes were unmistakably testes and ovaries always ovaries. It seemed the riddle had been solved, scientifically.

But nearly as soon as scientific precision inserted itself between parent and child, the monster of indeterminacy raised its seven heads. When scientists began to look more carefully through their microscopes at ovaries and testes, they soon found it wasn't always possible to lay those precious organs into one basket or the other. Sometimes gonads didn't really develop, and the examiners were left with only streaks of tissue—smears of cells that never quite organized themselves into functional organs—to characterize. Other times the gonads looked like both ovaries and testes, sometimes neither. As powerful as the sciences of anatomy and histology were, the false twins of sex disposed of them as quickly and as easily as it had blinded other human eyes.

Sex in the Genes

By the twentieth century, geneticists stepped forward to offer their help. It seemed to many that here, at last, was the answer—a candle to crack the dark. Karyotyping is an assessment of all of a person's chromosomes, a measure of each of the wormy curls that wrap themselves inside the nucleus of every one of our cells.

A karyotype requires only a few cells and the force to drive these cells into a state called "metaphase," where all the chromosomes become visible. A simple black and white photograph records everything for analysis. After that, all that is needed is a pair of sharp scissors and a person of great patience to cut out each of the little squiggles from the chromosomal picture and match them up with one another.

Usually, each of us has a total of forty-six chromosomes: twenty-two matched pairs, and two sex chromosomes that may or may not be matched. According to karyotypers, boys are 46,XY and girls are 46,XX—meaning "normal" girls have forty-six chromosomes including two X chromosomes, and "normal" boys have forty-six chromosomes including one X and one Y chromosome.

At first karyotyping seemed to have finally laid all ambiguity to rest. But pretty quickly we discovered that whole classes of people have more or fewer than forty-six chromosomes, often including unusual numbers of sex chromosomes. Even the geneticists had to admit that their central dogma, at least in its simplest form, had flaws. At the very least, though, it still seemed the basic tenet was true. It might be that a few folks had abnormal numbers of chromosomes, but if you had the requisite number plus an X and a Y you grew up to be a man. And if you had a set of forty-six that included two X chromosomes, you grew up to be a woman.

But even as karyotyping took hold in the late 1950s, both old and new studies came to light that cast some doubt on the absolutism of the genetics of sex. The first of these was a rediscovered study from 1945 that examined the lives of eighty-four hermaphrodites. These researchers concluded that "the hermaphrodite assumes a heterosexual libido and sex role that accords primarily with his or her masculine or physical upbringing," not his or her chromosomes.[21] A slight tremor rippled the ground beneath the geneticists' feet.

The second was a study of intersex patients conducted in 1955 at Johns Hopkins Hospital in Baltimore, Maryland.[22] This classic study further supported the earlier findings that "the gender assignment in infancy will be the one the patient stays with into adulthood, regardless of the standard biological indicators of sex." This meant that, to these researchers, it appeared that people tended to end up being what other people told them to be, not what their chromosomes predicted they should be. That fractured a lot of paradigms about childhood development.

The group at Johns Hopkins, as well as many others, dropped the "true sex" policy (such as XX or XY) and adopted the "optimal gender" policy for assigning sex to sexually ambiguous children.[23] According to this policy, proposed by Dr. John Money and his colleagues at Johns Hopkins, "the assignment or reassignment of gender should be based on the expected optimal outcome in terms of psychosexual, reproductive, and overall psychologic/psychiatric functioning."[24] Translated, that meant that it was up to the pediatricians, endocrinologists, plastic surgeons, and parents—not chromosomes, gonads, or the vagaries of biology—to determine the sex of the child. Furthermore, this protocol suggested surgery as early as possible to quickly unify physical appearance and the gender expectations of all involved. Though many of these ideas have since been discredited, they took hold like grappling hooks in the medical community of the time.

If a baby's sex wasn't obvious, as soon as possible those involved told the parents (if they wished to involve the parents at all) what sex would be best for their child. Then all moved quickly to create a physical reality for the child that is consistent with the chosen gender. The nature of sex had lost its prominence; in its place, the "optimal gender" policy raised up the force of nurture—the reality of sex by reason and design, sex by environment.

To this point, no one had considered the child's nervous system to be of any consequence in human sexual development. In infants especially, the mind and brain were thought to be of no consequence, but that perception was about to change. At the time, most people thought that at birth, and for the first few years afterward, the child was a blank slate. Masculine or feminine characteristics and perceptions could be pushed in either direction simply by manipulating the social, biological, and psychological fabric of a child's life. Like a neuter ball of clay, the child simply waited for hands and words to shape it into a boy or a girl.

But even before the 1950s ended, support for that view of sexuality began to fade. Near the end of that decade, a group of scientists studying guinea pigs discovered that early exposure to sex hormones had a major effect on sexual orientation of the brain.[25] Similar observations in several other species followed. Hormones, apparently, in addition to shaping our genitalia, also shaped our minds, and that happened perhaps in the first few weeks and certainly within the first months of life.

Soon after, another series of studies appeared showing that even prenatal exposure of 46,XX female infants to androgens (male sex hormones) produced girls and women with more masculine features and behavior.[26] This further bolstered the idea that a piece of the sex puzzle lay inside human brains, and that piece was highly responsive to the hormones that bathed the fetus.

In the early 1970s, our understanding of the course of human sexual development took an even more amazing and entirely unexpected turn. In 1971, a group of researchers described certain girls who, at puberty, seemed to become boys. An enzyme called 17-beta-hydroxysteroid dehydrogenase (17β-HSD) normally helps to convert androstenedione to testosterone in the developing fetus. Several different mutations can cause deficiencies in 17β-HSD. When that happens, there is less testosterone in the developing fetus. Because of that, most of these children—though 46,XY—are born with female genitalia and identified and raised as baby girls. But at puberty, many of these children begin to produce nearly male amounts of testosterone, develop phalluses much like penises, grow beards, acquire large muscles, and speak with the voices of men—they undergo a nearly complete reversal of sex.[27]

Then in 1974 another study appeared that described a second syndrome called 5-alpha reductase deficiency, or 5a-RD. This syndrome results from mutations in a gene encoding 5-alpha reductase (5aR). This enzyme is responsible for the production of another hormone essential for making baby boys. First, like boys with deficiencies in 17β-HSD, genetic males with 5α-RD are born with female external genitalia, and are consistently identified and raised as baby girls. And with 5α-RD, at puberty, these girls again become boys, or nearly so.[28]

Surprisingly, most children with deficiencies in either 5αR or 17β-HSD manage the gender switch at puberty without medication. This means, to many, that the gender identity of these boys results from the combined effects of prenatal and pubertal hormones and not the psychosexual environment in which the boys grew up.

Somewhere within that brew of hormones and genes and dolls and toy soldiers there may be some deep secret about how we all come to be who we are. Or maybe not. Regardless, two things are clear. First, no one has yet figured out the list of ingredients needed to make a boy or a girl—neither chromosomes, nor hormones, nor genes, nor family or society or chance, alone or in combination, seems sufficient to explain how one's sex comes to be. And second, there is no hard reason why we've come to believe that people even need to be a boy or a girl, no hard reason whatsoever.

The history of our wanderings in the maze of human sex makes for fascinating reading. But it does not make boys into boys, or girls into girls, or eliminate the gulf in between.

③

Sex Versus Reproduction: Why Are We So Married to the Idea of Two Sexes?

How we think about one another has clearly changed drastically over the millennia. But there has been one constant—we've always been thinking about it. Sexual intercourse, the force that drives the planet, has been a preoccupation of people forever. And one thing was clear from the outset: reproduction involves intercourse. From Hippocrates forward (and, I'm sure, backward), people have understood that people came from people, and the sex act was responsible for that.

But although it's clear that reproduction inevitably results from intercourse, the inverse is not true: sexual intercourse does not always result in reproduction. What, then, is its true purpose? Could it be that sexual intercourse serves other purposes? In a class I taught, we used a book with the question "Why sex?" written in large letters across the back. On her flight home at Thanksgiving, one of my students was reading that book when the woman sitting next to her noticed the question and said, "The answer is simple—because it feels so good."

Sexual interactions do feel good. They taste good. They even smell good. Indoors or out; in the morning, afternoon, or evening; missionary style or otherwise; in the shade or in the sun; sex catches all of our senses and holds them hard and fast. And when it's done with us, sex leaves us unclasping ourselves, struggling for breath, and weak as children. For most of us, nothing but sex can do that or anything even remotely like that. That feeling is what draws us and holds us, what pins us to one another for that long moment. It just feels *so* good. For most of us, that is reason enough for human sexual intercourse. But for biologists, that reason isn't quite enough. Somewhere in the warp of the cerebral cortices of these curious few, there remains an unanswered question: *Why* does sex feel so good?

After decades of cogitation, not to mention extensive experimentation, most biologists agree that sex feels so very good because sex creates more biologists. (Well, and others too, of course.) Which is to say that every human being alive today, as well as every nonhuman being, is here because every one of our direct ancestors—back to lizards and beyond—reproduced. All of them were drawn to sexual intercourse at least long enough to make one more like themselves, often many more like themselves. My mother, for example, came from a family of eleven; a sexual proclivity is definitely in my genes. It is in yours as well. Reproduction is the engine that drives all of biology. The more of you there are, the more successful you are, biologically. It's all about domination, and nothing dominates like sheer numbers. Reproduction provides the means; lust provides the motivation.

As a result, when asked "What is the purpose of sex?" most of us answer "reproduction." Why do we have penises and scrota, vaginas, clitorises, and uteruses? Again—for reproduction, almighty reproduction. This answer fits very nicely with our idea about two opposite sexes. But it also leads us into that dark labyrinth that opens with the idea that people without functional penises and vaginas and uteruses are abnormal because their genitalia aren't reproductive.

In reality, if you look closely, you'll find that in many species, including our own, reproduction is not the only aim of sexual interaction. And in some species, including our own, reproduction does not even appear to be the *primary* function of sexual interaction.

Among humans, in spite of our numbers, our sexual interactions are in fact rarely reproductive. First, most women are fertile about five or six days a month, or around 20 percent of the time. Most humans do not restrict their sexual activity to those five or six days. And in women there is nothing like the outwardly visible signs of "heat" that accompany female estrus in so many other species—nothing obvious to help focus human sex around periods of ovulation, to pull us together when we are most likely to produce a child. If the only purpose of sex is reproduction, that seems like a pretty major oversight on the part of evolution. Beyond that, even under ideal conditions, successful fertilization after intercourse occurs less than 50 percent of the time. After that, only about 50 percent of fertilized eggs ever develop beyond the blastula stage (reached about one week after fertilization). And only a small proportion of those blastulas successfully implant in the uterine wall. Then come all the dangers and pitfalls of development. Considering all of that, even by the most optimistic estimates, human sexual intercourse results in reproduction only about 5 percent of the time. Our phenomenal population growth rate is more a testament to our fondness for intercourse than it is to fecundity.

Beyond that, humans spend a lot of time having sexual intercourse when there is zero likelihood of reproduction. Between ages twenty-five and sixty, the percent of people having sex "at least a few times a month" only drops from 80 percent to 65 percent. But during that same period of our lives, essentially all women and many men go from fertile to sterile. It seems that the likelihood of reproduction has little to do with the magnetism of sexual intercourse.

Sexless Reproduction

If sheer numbers were the only measure of a species' success, then everybody on this planet should model themselves after bacteria. Bacteria are the most successful of all living species. Bacteria outweigh all other living things combined. They outnumber us by unimaginable factors. Bacteria are found in more places on the planet than any other living thing. There are more species of bacteria than all other species combined. And bacteria have had a greater impact on this planet than humans can ever hope to. Bacteria are number one. But bacteria reproduce by dividing—no sex, no lust, no titillation. Sexless cell division has pushed bacteria to the pinnacle of all biology. So it seems we, as a species, would be a lot better off if we reproduced asexually, like bacteria, periodically splitting off another one like ourselves to carry our genetic heritage into the future. Among species that reproduce asexually, every individual is female, and every individual can produce offspring. Taking their lead from bacteria, starfish do it, sponges do it, some worms do it, a lot of insects do it, and sometimes even turkeys do it. Somewhere around 1952, some folks in Beltsville, Maryland, noticed that about 16 percent of their turkey eggs developed, without fertilization, into brand-new hens. From those hens, the turkey ranchers developed a strain of turkeys whose eggs developed without fertilization about 40 percent of the time. No one expected that animals as complex and modern as turkeys could so easily be pushed into a single-sex way of life. By the criteria of sheer mass and numbers, most of these animals are much more success-ful than we are. So why do so many animals and so many plants devote so much time and energy to sex? Put most simply, we don't know. No one has ever unequivocally established why we have sex instead of just splitting in two every once in a while like bacteria.

The Selfish Gene and the Red Queen

The explanation of the purpose of sex offered most frequently is a theory called the "Red Queen" hypothesis, which is a simple corollary of the "selfish gene" hypothesis first proposed by Richard Dawkins.[1] Dawkins's hypothesis suggests that humans, as well as plants and other animals, are no more than containers built by evolution to move genes from place to place and to deliver those genes into spaces where they can make more copies of themselves.

At first that might seem absurd, but it has some merit. Imagine the beginnings of life in the warm seas of Earth. As the camera pans back, we see steamy seas surrounded by nothing but bare rock, some of it still molten. This is a bleak place and will remain so for billions of years. The camera then zooms in on the seawater and slips beneath the surface. Here the picture is only slightly different—a few dissolved chemicals and a lot more rock. But as we watch, a few of the chemicals snap themselves together like pop-beads and make long strings of molecules like DNA, molecules that look a little like genes. Once the pieces have strung themselves together, they acquire a most remarkable ability: from pieces floating in the nearby sea water these long molecules can make more like themselves, or very nearly like themselves. They can reproduce.

Abruptly, this chemical reaction spreads across the Earth's seas. Molecules are zipping themselves up everywhere, until only a few free pieces still float in the open water. Then, only the fastest of the big molecules can make more like themselves. The fastest molecules soon outrace the slower ones and dominate the early seas. But the raw materials continue to dwindle until even the fastest of these pre-genes have trouble making more copies of themselves. For a while, things nearly stop. Then one newly assembled large molecule discovers it can disassemble other large molecules and use the recycled materials for its own reproduction.

The process all starts over again, with the whole molecular world ripping away at itself until one molecule figures out how to wrap itself in an oil droplet to protect itself from the surrounding destruction. At that instant, the first cell is born. And all the rest spirals out from there—archaea, bacteria, plants, jellyfish, fish, lizards, humans—all simply more and more diverse vessels for the genes, vessels driven by genes, to protect genes, to transport genes, and to deliver genes into places where they can, just as they have for billions of years, make more copies of themselves. Selfish genes.

From time beyond recollection, genes have run everything. And, at least according to Dawkins, every act a human being has ever performed was dictated by one of those selfish genes. Writing poetry, for example.

The wind was a torrent of darkness upon the gusty trees,
The moon was a ghostly galleon tossed upon cloudy seas,
The road was a ribbon of moonlight looping the purple moor,
And the highwayman came riding—
    from "The Highwayman" by Alfred Noyes, 1907

Alfred Noyes may have seen his composition as a simple act of love for words and sounds, a little lust for the lyrical. To Richard Dawkins, though, Noyes was merely the pawn of his selfish genes, which manipulated Noyes for their own ends—reproduction. Noyes's intellectual intentions had nothing to do with the outcome of his work; his genes were just hoping to get him laid. Later that same year, Noyes married Garnett Daniels. Over the next few years, they had three children. And in 1925, Noyes converted to Catholicism, a calling known for its reproductive prowess.

Hard to argue with that—Alfred and Garnett and their visions of ghostly galleons, Richard Dawkins and his selfish genes—a DNA-driven plot to make more DNA. Better poetry through chemistry. The seductive selfish gene.

But if the selfish-gene theory is true, why bother with Alfred at all? Ms. Daniels, all on her own, could have produced many more children in the same amount of time had she been capable of parthenogenesis—asexual reproduction. If she could have done that, all the energy that Alfred poured into his poetry, all the food he consumed, the fires he burned, and the spaces he occupied would have been available to Ms. Daniels for making more women, who would make more women, and so on. Not only that, the fact remains that life forms that write terrible poetry (bacteria) and who know nothing about either sex or death remain the most successful of all living things on this planet.

So the question remains. Why do so many of us devote so much time and energy to sexual interaction?

In Lewis Carroll's *Through the Looking Glass and What Alice Found There*, Alice finds herself racing across a giant chessboard with the Red Queen. After a bit, Alice notices that even though she and the Queen are running as fast as they possibly can, everything around them is stationary. The trees, the flowers, the checkerboard—all appear to be standing still. The Queen finally tells Alice she can rest a while and props her against a tree. As Alice recovers her breath, she notices that this is the same tree where it all began. Speaking to the Queen, Alice says, "Why, I do believe we've been under this tree for the whole time! Everything is just as it was!"

To which the Red Queen replies, "Of course it is. What would you have it?"

"Well, in our country," says Alice, "you'd generally get to somewhere else—if you run very fast for a long time, as we've been doing."

"A slow sort of country!" says the Queen. "Now, here, you see, it takes all the running you can do, to keep in the same place. If you want to get somewhere else, you must run at least twice as fast as that!"

Our world is much like the world of the Red Queen. Every day of our lives, there are at least ten billion other living things that would like to eat us for lunch and claim the space that we currently occupy. The same is true for every other animal and plant on Earth. The proof of that appears whenever our defenses drop, for example, when an HIV infection strips one of us of an immune system and raises that thin curtain that stands between us and the rest of the world.

Then, in a relatively short period of time, where an individual had just stood, there is suddenly a community of living things—bacteria, viruses, fungi, and parasites. We exist only so long as we can defend our claims to our individuality and a small piece of the planet. Life is hard and tenuous.

To make matters even worse, our competitors—the microbes— never stand still. Every time our immune systems conjure up a new way to save us from one of them, the microbes come up with a new diversion of their own, a new way of fooling our defenses and slipping into our blood, or our brains, or our livers. The Red Queen hypothesis proposes that the only thing that saves us from this horrible fate is sex.

Bacteria divide once every twenty minutes or so, and at every division there is the opportunity for genetic change. In humans, it takes roughly twenty years. Clearly, bacteria evolve much faster than humans do. So our immune systems are always shooting at moving targets. Because of that, we must, as Alice and the Red Queen did, run as fast as we can just to stay in place. We must evolve as quickly as possible just to stay alive. If we didn't do that—if we didn't continuously reshuffle our genetic decks—human diversity would stagnate. If that happened, when one bacterium finally figured out a way around one human immune system, that bacterium might be able to get around all human immune systems. Then, if an infectious disease killed one man or woman, it might just as easily kill all of us.

The Red Queen theory says that our salvation is reproduction—the mixing of the genes that occurs every time sperm and ovum come together to start a new life. The more often we shuffle our genetic deck and deal ourselves new hands, the less likely it is that any microbial body will guess just which hand it must play to beat us. Without sex, our similarities would kill us.

The Red Queen theory seems sensible, but two rather large problems still remain. First, if sex is so critical to survival, why do we still have sponges and starfish and worms and insects and turkeys? It's not like these animals are struggling to make it; they outnumber us dramatically (except for turkeys). Then there are also all of those asexually reproducing plants that are doing just fine out there, thank you. Second, as discussed earlier, human sex seems in no particular way to focus on the act of reproduction. People have sex all of the time. What is that about?

Clearly reproduction requires sex. But maybe that is only one of many functions of this wondrous act. Is it possible that our views about sexuality and reproduction are merely a consequence of how we are brought up, the things we are taught to believe, an impact crater left by a religious meteorite? One way to answer that question is to take a look at our closest living relatives—relatives not burdened with our traditions, our mythologies, or our guilt, and see if they share our attitudes about sex.

Bonobos at Play

Because of our fascination with sex and our hubris about being the pinnacle of evolution, we have come to regard human sex as something sacred and unlike anything found elsewhere in the animal kingdom. In his landmark book *The Naked Ape*, zoologist Desmond Morris even went so far as to claim that humans are the sexiest primates alive, that women are the only females of any species to experience orgasms, and that men have the largest penises of all the primates.[2] Morris was wrong, on every count. With regard to each of these categories, a lesser-known primate, the bonobo, has men and women cleanly beat.

Bonobos are a remarkable group of apes found only in the lowland rainforests between the Congo and Kasi rivers in the Democratic Republic of Congo. Only a few groups of these apes exist, perhaps five thousand to twenty thousand individuals in all. Because of their dwindling numbers, the government of the Democratic Republic of Congo, with the help of the Bonobo Conservation Initiative in Washington, D.C., has recently moved to protect 30,500 square kilometers of bonobo habitat.

Bonobos were once and sometimes still are called "pygmy chimpanzees," which is not a very accurate name, since bonobos and chimps are nearly the same size, so "pygmy" doesn't fit, and in all other features bonobos differ markedly from chimps, so "chimpanzee" doesn't fit either. According to Frans de Waal—an expert on nonhuman primates in general and bonobos in particular—bonobos are sensitive, lively, and nervous, whereas chimpanzees are coarse and hot-tempered. Bonobos rarely raise their hair; chimpanzees often do. Physical violence almost never occurs among bonobos yet is common among chimpanzees. Bonobos defend themselves through aimed kicking with their feet, whereas chimpanzees pull their attackers close to bite them.

And, perhaps most interesting of all, bonobos copulate *more hominum* (face to face) and chimpanzees *more canum* (like dogs). The Latin was provided to protect the innocent, as well as the first writers of these findings, who knew that pointing out that bonobos had sex in this position long before humans did might have repercussions. After all, in the 1960s, cultural anthropologists had speculated at length about the importance of face-to-face mating and its elevation and sanctification of human lovemaking. Face-to-face sex became the infantization of lovemaking—with the male assuming the role of the infant and stimulating in the female the intimacies and kindnesses of motherhood. The female, too, became the suckling infant as she pressed against the male's chest and prepared to receive "a life-giving liquid from an adult bodily protuberance."[3]

Besides their fondness for human-like lovemaking, bonobos just seem a lot kinder and gentler than their chimpanzee cousins. Genetically, bonobos are about 98 percent identical to humans—a little more like us than chimps are. In captivity, bonobos have lived as long as fifty years. The males are about two-and-one-half feet tall and weigh between seventy-five and one hundred pounds. The females are only a little smaller. Both male and female bonobos often walk upright, just as we do. Chimps do so much less often. All of that—especially their genetic makeup—makes bonobos look like our closest living relatives. And our closest living relatives, it turns out, devote more time, energy, and creativity to sexual interactions—in all forms—than do any other nonhuman primates. And among their purposes for sex, reproduction appears to fall fairly low on the list of priorities.

In fact, the creativity and proclivity shown by bonobos during their sexual interactions are truly astonishing. "The best way to convey the richness of this ape's sexuality," wrote de Waal, "is to list the patterns observed at the San Diego Zoo. Before I went there, I had heard that the bonobos were sexy, but I was nonetheless amazed by the sheer variety of positions and the extent to which the apes mutually stimulated each other."[4]

In spite of his extensive knowledge of apes, what de Waal saw at the San Diego Zoo still stunned him. He goes on to describe a variety of interactions that could exhaust even the most libidinous among us.

More canum, or dog-style intercourse between males and females, is the most common sort of sexual interaction observed among the bonobos, as it is with chimpanzees. This is somewhat surprising because bonobos, unlike chimps, share with us forward-facing genitals, which seem best suited for face-to-face sex. The second most common interaction was missionary style, which seems to be the favorite among female bonobos. As sexual arousal overtakes a female bonobo, her labia swell to the size of grapefruits. At the same time her clitoris becomes prominently erectile and protrudes toward the front. When she initiates sex, she is almost always lying on her back. And even when a male tries to initiate sex from a different position she may push him into the position she prefers, the position where she receives the most clitoral stimulation during intercourse.

Still, dog-style couplings occur about twice as often as human-style couplings among bonobos. Beyond that, there is a virtual cornucopia of sexual activities. For example, dalliances are also common between two females. In this interaction, one female lifts the other into her arms and carries her like she might a child. Arranged like this, the two females can and do rub their clitorises together with exactly the same frequency that males thrust into females during copulation. Scientists who study bonobos call this genito-genital rubbing, or GG rubbing, and among the apes, only bonobos do it, and they do it a lot.

Other times, two female bonobos will face opposite directions, one lying on her back with her legs spread, the other backing up to her and squatting until their genitals make contact, and then they stimulate one another with sideways movements of the genitalia.

Not to be outdone, males engage in similar sorts of contact. Both males remain on all fours and back up against one another until their rumps and their scrota come together. This action, called rump-rump contact, is usually brief, but clearly pleasant for both partners.

Males also have sex facing one another. One of the partners lies on his back, and the other, usually older, partner straddles him until their penises come together. Both males have erections, and the older one thrusts his penis rhythmically against the other male's penis. De Waal says that he has never seen either of the males in this sort of interaction achieve orgasm. Nor has he ever seen any attempt at anal intercourse.

Others have also described "penis fencing" between male bonobos—a much more rarely observed activity in which two males hang from a tree branch facing one another and rub their erect penises together like fencing swords.

These activities are not restricted to adults. Juvenile bonobos also often participate. De Waal describes one infant female who would often jump on top of adults engaged in intercourse, sometimes press her genitalia against her mother's while her mother was engaged in GG rubbing with another adult female, and would climb on top of adolescent males and press herself against their penises. The males would usually respond with a few thrusts but no penetration or orgasm.

Bonobos also kiss, face-to-face, open-mouthed, tongues fully entwined. Chimpanzee kisses, on the other hand, tend to be rather perfunctory in nature. Chimpanzees never French kiss. According to de Waal, "this explains why a new zoo keeper [at the San Diego Zoo] familiar with chimpanzees once accepted a kiss from a bonobo. Was he taken aback when he suddenly felt the ape's tongue inside his mouth!"

Bonobos have a fondness for fellatio—one male taking the erect penis of another into his mouth to provide pleasure. Fellatio happens frequently during juvenile play among male bonobos.

Masturbation also plays a big role in bonobos' lives—both solitary masturbation and masturbation with partners. Often a young male will present his erect penis to an adult male who will "loosely close his hand around the shaft, making caressing up-and-down movements." In addition, both sexes engage in solitary masturbation. The most regular participants are young males and adult females.

Here again, it would seem that Desmond Morris's claims for the supremacy of human sex and for the sole female orgasm of all the species in the world rings hollow. For, as de Waal asks, why would female bonobos masturbate if not for pleasure? Beyond that, studies by de Waal and others show that physiology and behavior of female bonobos during intercourse and masturbation are consistent with orgasms.[5]

Sex Before Surrender

But there is more to bonobo sex than play and pleasure. For most primates, life is competition, and the outcomes of their competitions often determine the length and quality of primates' lives. One of the most routine of these competitions involves the fight for food. Who gets to eat first, and who gets the most food? Often physical violence or the threat of physical violence—dominance of one ape over another—resolves these food issues. Bonobos, however, seem to have found another way.

According to de Waal, when zookeepers approach bonobos with food, the males sprout erections. And before the food even hits the floor of their enclosure, the bonobos start thinking about sex. Males engage females, females seduce males, other females are GG rubbing, and a near orgy ensues. No one fully understands why bonobos exhibit this sexual frenzy, but experts like de Waal have suggested some possibilities. He suggests that, in this instance, sex eases social tension and is a surrogate for confrontation and fighting, which, as he points out, is not unheard of in other species, including humans.

Competition for food is not the only conflict that triggers bonobo sex. An encounter with a cardboard box or even a simple unexpected piece of rope may be enough. When two bonobos spot the box or rope, a little lovemaking or GG rubbing often precedes exploration of the item itself. And bonobo females sometimes trade sex for food. De Waal recalls a time he spotted "a young female grinning and squealing during copulation with a male who held two oranges, one in each hand. The female had presented herself to him as soon as she noticed what he had. She walked away from the scene with one of the two fruits."

De Waal offers a few other thoughts about why it may be so important to bonobos and other primates that their interest in and availability for sex is not limited to periods of fertility, as it is in dogs and cats and horses and cows. Many primates—especially humans, bonobos, and chimps—are born nearly helpless. Survival to adulthood depends very much on the nuclear family—mothers and fathers nurse, care for, and protect the babies. The permanence of that family bond is essential to our survival. Perhaps bonobos' and humans' interest in sex also serves the purpose of gluing us to one another, of stabilizing the nuclear unit for the well-being of our babies.

The Truth About Human Sex?

Sex as solution, greeting, diversion, alternative to aggression, appeasement, social currency, curiosity, collateral, welcome, wonder, and waltz—clearly, sexual relations play a rich and complex series of roles in bonobo society and raise important questions about our own views on the purpose of sex. It almost seems that reproduction is a nearly accidental by-product of all the other ways in which sex serves bonobos. Maybe among certain species, especially some primates, sex has evolved beyond its simple reproductive beginnings. Perhaps sex once served the sole purpose of reproduction, but over millions of years has evolved—just as eyes and ears and muscles and fingers evolved—to serve a multitude of more complex functions, only one of which still focuses on reproduction. We humans even go so far as to actively subvert the reproductive function of sex in favor of all the other things that sex does for us. If the sole purpose of sex were to reproduce ourselves, birth control would seem evolutionary heresy.

Any species seeking to sate its desires, rework its social structure, and avoid confrontation would set itself on a path to oblivion if the sole purpose of sex were reproduction. The fact that we and the bonobos are still here is testament to the fact that as we evolved we co-opted the sexual urge for a multitude of other purposes—purposes essential to our past, our present, and our future. These purposes are essential to the growth and development of our children—purposes of peace, cooperation, entrepreneurship, pleasure, and promise. If that isn't sufficient reason for all of us to reconsider how we think about ourselves and our sexes, then at the very least it should be cause for reflection on the value and judgments we place on people whose genitalia don't obviously lend themselves to reproduction.

4

WHERE OUR SEXES COME FROM:
THE ABRIDGED VERSION

"I am afraid we could lose them both," the old doctor whispered through his clenched teeth as he pushed his hand through his thin hair and watched the woman writhing in pain.

One of the baby's shoulders was stuck inside the mother's pelvic girdle. The doctor, an old-fashioned general practitioner, had no idea what to do next. He had never seen anything like this. As his ignorance enveloped him, he became a spectator. The father, a youngish man with dark black hair and fearful eyes, struggled to keep the ether-soaked cotton sponge near the mother's nose—near enough to ease the pain, but not so close as to nudge her into unconsciousness. A petroleum engineer by trade, just now the father was an anesthesiologist.

Half in and half out, the child gasped for breath as the contractions continued to rip through the mother's womb. She groaned loudly with each new spasm. For another minute, which seemed like an hour, things stayed just as they were—baby stuck between worlds, mother in agony, and father and doctor fearing the worst.

Then, with a deep shrug and a heavy sigh, I was abruptly spit into this world of light and cold and fear and beauty. I opened my eyes wide, took one long look at everything and everybody in that room, and screamed.

My mother exhaled, my father dropped his ether rag, and a smile spread like a warm fire across the doctor's face.

"What is it?" my mother asked.

"Why, it's a boy," the doctor said.

My mother smiled broadly. She might have been less pleased had she known that the doctor based his statement on a surprisingly small amount of evidence—tiny bits of flesh that looked like penis and scrotum but could have been any number of other things. The doctor, with only one alternative, had simply settled with the only story he knew about life's beginnings.

As it turns out, the story of boy versus girl versus everything else is much more complicated than the doctor knew, more complicated than any of us knew, perhaps more complicated than any of us can know.

The Story of Life and Sex

When we tell the story of how a human baby came to be, we usually begin at the moment when sperm meets egg. Maybe that's because we generally don't think of eggs and sperm as human beings. Eggs carry most of what it takes to make a human being. In fact, eggs contain everything needed to make another human being except for twenty-three chromosomes. Those twenty-three chromosomes come from a sperm, and that is pretty much all that the sperm brings to this blessed event.

A human egg begins its life in an ovary as the end result of a mysterious process called meiosis. Meiosis is a peculiar type of cell division that eventually produces cells with twenty-three chromosomes. Twenty-three is one-half the number of chromosomes found in all the rest of the cells inside of a human being. So we call sperm and ova haploid cells—half cells—and it takes two cells (egg and sperm) to create a complete human being whose cells each have forty-six chromosomes (diploid cells).

About once a month in a sexually mature woman, one of the haploid eggs matures, slips into the Fallopian tube, and heads off toward the uterus—a relatively short journey, distance-wise, but a journey pregnant with possibilities. If the egg never encounters a sperm, then this ovulation leads to menstruation as the uterus sloughs the bed it had prepared for an embryo.

If, on the other hand, sperm find their way into the vagina during coitus at the right time, the events that follow may be entirely different. Sperm are fertilization machines—twenty-three chromosomes and an eggbeater for a tail. Sperm exist for a single purpose—finding and fertilizing eggs; it is their raison d'etre.

From the vagina, sperm head for the uterus, and from the uterus to the Fallopian tubes. Inside these moist tunnels, fertilization becomes possible. The first sperm to reach its goal augers into the egg and slams the door behind it, usually guaranteeing that only a single sperm will fertilize a single egg. Then things get really intense. The chromosomes carried by egg and sperm gather together, and a single diploid cell appears, a single cell from which all others will arise—the bones and tendons of the foot, the neurons of the amygdala, the lens of the eye, the crease of the lips, the fertile ponds of the genitals, the curious curl of the iris, the whorl of a fingerprint.

Imagine the complexity of that process, which will produce hundreds of billions of cells, no two exactly alike, cells capable of all the things that human beings have ever done or thought, cells that can reach out or withhold, cells to listen with, cells to see with, cells to touch one another at the instant we most need to be touched, cells each more intricate than the most complex human creation. Imagine the possibilities for mistakes. During fetal development the opportunities for error, and the consequence for those errors, are as great as they will ever be. The leading cause of infant mortality is still developmental abnormalities—mistakes made along the way. And that way is so full of twists and turns and blind alleys that we know more about the surface of the moon than we truly understand about what drives and directs embryonic development. Yet that process affects nearly everything about us—our eyes, our ears, our feelings, our sanity, our toes, our bottoms, and our tops. And our sexes.

In fact, in biologists' minds, maybe in most of our minds, the process of embryonic development directly, inevitably, solely, and finally determines our sex. In fact, the one aspect of human development about which most people would agree that nurture plays no role whatsoever is human sex. Sex—boy versus girl—is all nature, we believe. In perhaps no other aspect of human development would we so quickly conclude that genes and chromosomes alone are solely responsible for a human trait. Even with things like eye color and hair color we anticipate environmental effects and changes during a person's lifetime. If you have two X chromosomes, you are first a girl and then a woman. If you have an X chromosome and a Y chromosome, you are first a boy and then a man—period.

Why we even believe such a tale about how sex is created is a mystery. We don't think that way about our personal habits, muscles, height, or even sanity. We reserve very special, elaborate, carefully crafted closets inside our minds for thoughts about sex. And inside of those closets we have rules, rules that no one can break without consequence. For that reason it is worth recapitulating this story we most like to tell about how we become boys or girls—forgetting for the moment, as we so often do, those who never do become boys or girls.

Before the newly formed embryo even leaves the Fallopian tubes, the cells that will become the gonads have separated from the others. Through the first four days and five rounds of cell division—until roughly the thirty-two-cell stage—the embryo grows as a solid ball of cells, a clump of tightly clustered grapes. Then, following some direction laid down a hundred million years ago, a few cells condense into a mass on one side of the cluster, and a fluid-filled cavity opens at the opposite side. Surrounding all of this is a thin crystal-clear membrane, a shell.

At the end of the fifth day, the embryo begins to rhythmically expand and contract. As those contractions intensify, the shell bursts, and the embryo hatches. All of this happens as the embryo slowly wanders through the labyrinth of the Fallopian tube. But by about day six this odyssey comes to an end, and the embryo implants itself in the frothy wall of the uterus.

The embryo now has two inner cavities—one will become the amniotic sac that will surround the developing fetus, and the other the secondary yolk sac. In between those two cavities is a thin strip of cells.

Within that narrow thread live the cells that will become the child—the living bridge between placenta and yolk, a bridge between amoeba and ape, between an exploding universe and the darknesses still to come.

Out of that cyclone of dividing cells, one of the first groups of cells to commit itself to a final future is the group that will form the gonads. These cells may become ovaries inside a new baby girl or testes in a new baby boy—bits of life destined by the finger of a force beyond our imagining to generate eggs or sperm. And all of that happens before we are even four weeks old.

By about week six, the embryo has assumed a vaguely recognizable form, and at a point just below the umbilical attachment the germ cells have settled into a structure called the gonadal or urogenital ridge. If some mutation prevents the formation of the urogenital ridge, no gonads and no sexual characteristics will ever develop, and usually no kidneys or adrenal glands will develop either. In this case, the embryo turns out its small lamps and slips back beneath its mother's blankets; it stops developing, and the uterus reabsorbs it.

But mutations that prevent the formation of the urogenital ridge are rare, which seems nearly a miracle, because, along the way to the gonadal ridge, dozens, perhaps hundreds of things need to occur with Swiss precision—enzymes must appear just as they are needed, then just as quickly fade. Pieces of DNA must be awakened at a precise hour, then lulled again to sleep. RNA must ratchet through the embryo's world, grinding out proteins like cars off an assembly line. Groups of cells must migrate from south to north, east to west, and back again.

Like an orchestra, each section plays its part—the reeds, the brass, the percussion, and the strings all make precisely timed and critical contributions. Occasionally in the brass section, French horns are swapped for flutes, and among the violins, snare drums appear; new musicians arrive and depart, the floor raises and lowers periodically, as does the curtain. And without any one piece of this, things change, the music shifts, often in unpredictable ways.

But if everything goes more or less as usual, at about seven weeks a pair of gonads appears near the urogenital ridge—gonads already filled with cells that will one day become eggs or sperm. At this point the gonads are somewhere between male and female; the final sex of the fetus depends on what happens next. If nothing (or nearly nothing) happens, the primitive gonads become ovaries, and the embryo develops the external and internal machinery of a girl—ovaries, uterus, Fallopian tubes, vagina, clitoris, labia. Female, in other words, is development's default mode; without orders to the contrary, women's bodies make more women. Perhaps that isn't as surprising as it might seem, given that sponges, starfish, whiptail lizards, and others do the same thing—all without males. Since we evolved from one of those all-female species, maybe it isn't startling that reproduction's first choice is female. Only the gradual evolution of the Y chromosome (from an X chromosome) changed that.

Compared to making girls, making boys is harder work. And though many things can go wrong on the way to making girl babies, there are even more potential detours, pitfalls, and wrong turns on the road to making baby boys.

The human X chromosome is long, wasp-waisted, and full of information. The Y chromosome, on the other hand, is an odd-looking little blob of DNA. But there are genes on the Y chromosome that can make all the difference when it comes to determining sex. One part in particular of the Y chromosome plays a critical role. We call it the sex-determining region of the Y chromosome, or SRY, and it is essential for making baby boys.

Genes in this part of the Y chromosome operate like switches to turn on other genes that are critical to male development. Without SRY, the primitive gonads do not develop into testes, and sometimes don't develop at all.[1] With SRY, the gonads, until now amorphous little tissues, transform into testes and by about four months begin to do some of the things that testes do.

The SRY part of the Y chromosome is essential for making baby boys, but it is not sufficient. After the testes form, there is another gauntlet that the developing fetus must run.

Now is the time for hormones. All fetuses produce about the same amount of estrogen, and if no other hormones appear, baby girls follow. But if the testes do what they normally do, baby boys follow.

In the eighty-day-old fetus, neurons in the fetal hypothalamus (a knot of cells at the base of the brain) begin to secrete pulses of gonado-tropin-releasing hormone (GNRH). In response, the fetal pituitary gland (another clump of brain cells that sits just below the hypothalamus) begins to release pulses of follicle-stimulating hormone and luteinizing hormone. These hormones stimulate the developing gonads (either tes-tes or ovaries), which in turn respond with hormones of their own.

For a developing fetus to end up looking like a baby boy two things have to happen: the testes must produce testosterone, and some of that testosterone must become 5-alpha dihydrotestosterone, another male hormone that helps steer fetal development. That conversion requires an enzyme called 5-alpha reductase. 5-alpha reductase comes in two forms, poetically named 5-alpha reductase types 1 and 2. Toward the end of fetal development, the appearance of testosterone, 5-alpha reductase type 2, and 5-alpha dihydrotestosterone sculpt the fetus into a baby boy with a penis, scrotum, and all the rest of the external and internal accoutrements of a male human being.

And since all hormones do their work through specific receptors on the developing child's cells, not only do all of the genes that direct hormone formation have to work in a timely and accurate manner, but so do all the genes that encode the cell-surface receptors for all of those hormones. And when each hormone delivers its message to its person-alized receptor on a cell's surface, then a whole series of other molecules have to tell the nucleus about that message before any critical changes in that cell's life can occur. So every one of those message-carrying mol-ecules has to be in place, on time, and ready to go or else everything produced by the testes or the developing adrenal glands will have no effect on anything.

That's a lot—a lot of molecules, a lot of genes, a lot of coordina-tion—that has to occur properly. And all of it happens during the first ten weeks of pregnancy. At ten weeks, you still can't tell a male from a female fetus. Even though they both have genitalia at this point, there are no differences—the cells that will become either the glans of the penis or the clitoris sit atop a short stalk that will transform into the shaft of the penis in males and into the folds of the vaginal opening in females. The whole fetus is just a little more than an inch long at this point. The tissues at the base of the stalk hold the still-developing gonads and will usually differentiate into either a scrotum or the tissues that surround and buttress the vulva.

While the external genitalia are still struggling to distinguish themselves, at ten weeks the internal genitalia begin to wander down separate paths. In developing females, the uterus, vagina, Fallopian tubes, and urethra assemble themselves and assume their proper positions. Similarly, in developing male fetuses the vas deferens, seminal vesicles, seminiferous tubules, prostate, epididymis, and ejaculatory orifice all form from the surrounding tissues. That's a lot of tubes and orifices, folds and bumps, hollows, vaults, and vesicles to be built and slipped into position. If the fetal gonads have been pumping out androgens during the first twelve weeks of fetal life, openings close, the scrotum forms, and the developing urethra finds its way to the tip of the penis. If no testosterone shows up, then the glans shrinks into a clitoris, the vagina fully opens, labia blossom, and the urethra ends up below the clitoris. A few long solo notes by the gonads, and the deal is sealed.

At about thirty-six weeks in developing boys, the testes descend into the scrotum. When that happens, the fetus finally has external genitalia that are fully male. In developing girls, at about the same time, the finishing touches appear, the excesses get wiped away, and the final version of the genitalia is formed.

Then the curtain lifts, the conductor raises a hand, a low tune begins among the bassoons, and a hush sweeps through the audience. A child is born.

More than any other single factor, the external genitalia, those few simple tissues, their sizes and their shapes, will imprint in everyone's mind an everlasting label of girl or boy, or neither—for some parents the worst possible outcome.

In spite of the label the doctor slaps on us at birth like a bar code, this whole sex process isn't over, not nearly. The pulses of hypothalamic, pituitary, and sex hormones that began in the eighty-day-old fetus and continued through the last two trimesters of fetal life continue still throughout early infancy. Then, at about six months of age, everything shuts down for a bit. This is called the "juvenile pause," when the hypothalamus, pituitary, and gonads put their feet up and take a well-deserved break.[2] The hormone storm abates, bones grow, muscles strengthen, brains wire and rewire themselves. But the sexual cyclone is still on course, and in most of us it will make landfall by about age twelve.

Shortly before the final stage of puberty, neurons within our hypothalamuses rise from their slumber and, while the rest of our bodies sleep, again begin to pump GNRH hormone into our pituitaries. Our pituitaries respond by releasing follicle-stimulating and luteinizing hormones. Then, every night during sleep, about every ninety minutes our gonads get a jolt of hormones. Ovaries begin to release estrogen, testes create testosterone and dihydrotestosterone, and the shapes of our bodies begin to change.

Swelling of one or both areolae (the circle around the nipples) and the first appearance of pubic hair along the labia mark the first brushstrokes of puberty in girls. Six to twelve months later that swelling has spread to both breasts and reaches beyond the areolae. At the same time, the pubic hair spreads to the pubic mound. About a year later, breasts have enlarged to nearly final size, and pubic hair has assumed the adult triangle-shaped appearance. Internally, another map is unfolding. The vagina, uterus, and ovaries all respond to this new spate of hormones as well. All enlarge and reach their final shapes and sizes. And, most of the time, ovulation and menstruation begin.

In infants with one X and one Y chromosome, usually the penis that formed before birth enlarges during the first four years of infancy, and then rests. At about age eleven, the testes begin to produce testosterone, the penis and scrotum enlarge, and pubic hair begins to appear. Ejaculation becomes possible early in puberty though ejaculates may contain relatively few sperm. At about the same time, the penis enlarges, and pubic hair becomes more abundant. About twelve months later the adult, triangle-shaped growth of pubic hair appears. And in the last stages, pubic hair spreads to the thighs and abdomen, facial hair begins to grow, and voices deepen.

Most of the time.

And most of the time all of these things happen because of those once-every-ninety-minute pulses of hormones put out by our hypo-thalamuses—small pistons compressing and relaxing, driven by their own biorhythms, pushing us, while we sleep, into adulthood. In response to these periodic squirts of GNRH, our pituitaries begin a cycle of on-off production of luteinizing and follicle-stimulating hormones. These hormones, in turn, seek out our testes or our ovaries and push them to secrete the all-important sex steroid hormones, estradiol and testosterone.

In men, at the same time, another form of the enzyme 5-alpha reductase type 1 appears. This enzyme, much like the one produced during fetal life, converts testosterone into 5-alpha dihydrotestosterone, another key needed to open the last gates on the way to male development.

Again, the entire orchestra must contribute. If even a single instrument fails to deliver its part on time and in proper syncopation, the whole symphony may crash. Hormones, enzymes, genes, neurons, receptors, and signal transduction machinery can shatter into a million shards, and the images those shards reflect can be as varied as our imaginations.

And as complex as all that may seem, it's only part of the story. The adrenal glands are another set of critical organs that form with the genital ridge as it develops in fetuses. The adrenal glands also play critical roles in the process of human sexual development. In response to hormones produced by the hypothalamus and the pituitary glands during fetal development and during puberty, the adrenal glands typically make and secrete several additional hormones. Among these are the adrenal androgens—dehydroepiandrosterone (DHEA), androstene-dione, and pregnenolone. These compounds by themselves don't do very much. But in the blood, enzymes convert these precursors into the active hormones testosterone, dihydrotestosterone, and progester-one—hormones critical to our initial and continued development as complex beings.[3]

The exact role of androgens in female development is cloudy, but many aspects of normal female development—including nerve and muscle cell development, as well as hormone production—do not occur normally in women with adrenal androgen deficiencies.[4]

Boys begin to produce DHEA at about age nine or ten and andro-stenedione a year or two later. With boys the adrenal androgens induce underarm hair and some pubic hair, but do not appear to be involved with other aspects of puberty. From birth on, the adrenal androgens act like another set of cogs in the machine that drives male sexual develop-ment. And, as we will see later, when these adrenal androgens become estrogens, which they can and sometimes do, men begin to look more like women.

And there are many more changes to come. Sexual development is a lifelong journey for us all. And though there are some common signposts along the way, such as puberty and menopause, we shall see how the journey is shrouded in mist, and the path is as varied as human beings themselves.

5

WHERE OUR SEXES COME FROM: THE REST OF THE STORY

That is the story of how we come to be—the routes we follow, the stops we make, and the forests and rivers and valleys we pass along the way. It's more or less everyone's story about the road from sperm and ovum to boy- or girlhood. Or so we believe, until we look a little more carefully at the map.

There are more than a million ways to get from Chicago to Los Angeles. But the route that has been written and sung about most is Route 66—that fine old stretch of macadam that runs across half a continent. But even those who follow the same highway see different things along the way, take different side trips, break down in different places, drink different coffee, and eat different pies. Some people stop now and then, by choice or by necessity. Some even die inside lonely motels left like droppings alongside the road. And though the ones who make it all the way all end up in Los Angeles, by the time they arrive no two of them are alike.

But that's not the story we generally tell. The story we generally tell is a much simpler version, a condensed tale that speaks only of the wonders along the way and what a great time all of the travelers had as they sped westward. The real story is a lot more complicated, the real ride a great deal more dangerous, the actual experience a good deal more nerve-wracking.

The simplest stories are rarely the truest.

The same can be said about sex. Most of the tales we spin about human sex determinations leave out the busted axle in Baxter Springs, the blown radiator in Barstow, the week spent washing dishes in Tucumcari for gas money. Most of the tales we tell about sex speak only a little of the road from conception to birth and instead move quickly into the safety of a pink or blue fog.

The shame in that is that this is *our* story, and the complete tale is full of wonders.

The Truth about X and Y

"OK, how do you know what sex you are, or what sex anyone is, for that matter?" I ask my class.

Alicia, a bright young woman majoring in biochemistry, raises her hand.

"Yes?"

"Chromosomes," she says. "Men are XY and women are XX."

"But wait a minute," says Clarisse, a liberal arts major. "What about XXY people? Or XO?"

Another student raises his hand and says, "Aren't there some women who are XY?"

And from there the discussion descends into chaos.

In spite of the confusion, though, it still seems that sex should have something to do with chromosomes. In fact it does, at least sometimes, but not how most of us imagine it does.

Lying like gene-stuffed serpents inside each of our cells, chromosomes seem to have acquired mysterious powers. Inside those coils of DNA, geneticists have told us, is our future—the color of our hair and eyes, the length of our bones, the size of our brains, the paths of our lives, and most directly and importantly, our sexes. But before a chromosome can do anything except sit around and soak up mutations, it needs aid, usually a protein. Proteins open chromosomes up and close them down, turn genes on or off, build a cell up or break it down. Every step on the way from zygote to adult is directed by one or more proteins. And each protein has its own set of problems and its own peculiar way of dealing with the world.

That's a little like having a set of blueprints that change in critical ways every time someone opens them up for guidance. And it's a lot like having carpenters, electricians, plumbers, and contractors who all speak slightly different languages. So, even if the blueprints were perfect to begin with (which they never are), the project's possible outcomes are as variable as the people who use those blueprints, as convoluted as spoken languages, as uncertain as a lottery.

It's the same with sex, but even more so.

Sex Chromosomes Lost and Found

Contrary to what biology texts have alleged as law, some people—genuine human being people—don't have forty-six chromosomes.

People with Down syndrome, for example, have forty-seven chromosomes, including an extra copy of chromosome twenty-one. That extra chromosome leaves its mark on these people—epicanthic folds in their eyelids, the slant to their eyes, simian creases in their palms, the often-associated mental retardation, their unabashed beauty.

There is a moment in most every cell's life when it imagines that it contains enough of everything for two. As that idea blossoms, the cell gathers together all of its chromosomes, waiting for the universe's own hand to split them apart and deliver one into each of the new cells. On the road to sperm or egg, in the final division of meiosis, the chromosomes that came from our mothers and our fathers gather, singly. Usually, as the cell pulls itself in two, one chromosome lands inside each of the daughter cells with perfect parity, but not always. Sometimes chromosomes cling to one another, and stick. Then, when the dividing cell tries to pull them apart, the chromosomes don't budge. A brief tug-of-war ensues, then one side gives in, and both chromosomes fall into only one of the daughter cells. The other cell gets none.

After that, when a sperm or an egg with two copies of chromosome

21 combines with a sperm or an egg containing the usual one copy of chromosome 21, a baby begins, and each of his or her cells has three copies of chromosome 21—trisomy 21.

We call this nondisjunction—chromosomes that fail to disjoin as the mother cell divides. And we call these people abnormal, because each of their cells holds one extra strand of DNA, because they have forty-seven instead of forty-six chromosomes.

Nondisjunction can happen with any chromosome, including the sex chromosomes X and Y. A single sperm or egg may end up with two, three, or more X chromosomes, and a single sperm may hold more than one Y chromosome. In truth, sperm and eggs come in variety packs.

If that alone isn't enough to derail the simple XX/XY, female/male idea, a mystery known as anaphase lag can also cause developing sperm or ova to lose an X or a Y chromosome along the way. And even after fertilization, sex chromosomes can be lost or gained. And even among men with the normal 46,XY karyotype, the size of the Y chromosome can vary.[1] That means that my Y chromosome might be three times the size of Arnold Schwarzenegger's Y chromosome. Here certainly, quantity matters; perhaps size does as well.

The end product is a panoply of possible sexes by any definition, an array of human beings as grand and as varietal as the fragrances of flowers: 45,X; 47,XXX; 48,XXXX; 49,XXXXX; 47,XYY; 47,XXY; 48,XXXY; 49,XXXXY; and 49,XXXYY.

People as Syndromes: Dr. Klinefelter's Discovery

In 1937 Harry Fitch Klinefelter graduated from Johns Hopkins Medical School and took a job at Harvard Medical School working with Fuller Albright, a physician famous for his work in endocrinology. Among Dr. Klinefelter's first patients was an African American man with enlarged breasts and small testes—something the doctor had not seen before. But the more he searched, the more patients like this he found. In 1942 Dr. Klinefelter published a paper describing nine such men. Each was sterile, each had gynecomastia (enlarged breasts), unusually long arms, normal-sized penises, and small testes.[2]

It turned out that each of these men also had forty-seven chromosomes. Each was 47,XXY. This syndrome became known as Klinefelter syndrome and grew to include all karyotypes with more than one X chromosome plus one or two Y chromosomes. Because of the presence of a Y chromosome (containing SRY), all of these people have penises. Because of that we call them males, but they all have more than their fair share of X chromosomes.

Only about 40 percent of fetuses with extra X chromosomes survive to term. Nevertheless, somewhere between 1 in 500 and 1 in 1,000 boys born in any given day have one or more extra X chromosomes.[3] That makes Klinefelter syndrome the most numerous among those syndromes with an unusual number of chromosomes.

In the end, this adds up to about three thousand live births every year in the United States, and maybe as many as sixty thousand worldwide, of boys with extra X chromosomes. Most of these men are never diagnosed. Nothing about these people would allow anyone except a geneticist to even suspect that they differ from others. Of those diagnosed, most find out only when they marry, fail to have children, and then undergo sterility testing.

The least common form of Klinefelter syndrome is the 48,XXYY karyotype. These people very closely resemble those with the 47,XXY karyotype except 48,XXYY men are taller, nearly five inches taller, on average, than their fathers. Apparently, one of the things Y chromosomes do is provide the necessary genetic material for the tallness of men in comparison to women. And in a double dose, it seems to push people a little higher.

Dr. Turner's Findings

The second largest group with unusual numbers of chromosomes includes those with the 45,X karyotype, commonly called XO or Turner syndrome.

When a sperm containing an X chromosome fertilizes an ovum with no X chromosome, the zygote that forms has only forty-five chromosomes. Or if a chromosome gets lost because of anaphase lag or some other reason, the zygote that develops will have only forty-five chromosomes, 45,X.

Since Henry Turner first described some of these people in 1938, we call it Turner syndrome. But of course it's more than a syndrome, much more. Just as with every other differently numbered collection of chromosomes, for those who survive it, 45,X is a way of life.

The 45,X karyotype appears in about 0.8 percent of zygotes. That makes 45,X the most common human chromosomal anomaly. But only about 3 percent of these fetuses survive to term. 45,X is also the most common chromosomal anomaly found in spontaneously aborted human fetuses—nearly 18 percent.[4] In the end, about one in 2,700 live newborns has the 45,X karyotype.[5] That is about fifteen hundred newborns per year in the United States and perhaps as many as thirty thousand worldwide.[6]

Just like the extra X chromosome, the absent X chromosome changes things. Externally and internally, at birth these people are female, although they are not identical to 46,XX females. The external genitalia are those of a girl. The vagina and uterus usually develop normally as well, but the ovaries do not. In place of ovaries, 45,X people have only streaks of tissue called streak gonads, or sometimes streak ovaries. Streak gonads do not produce normal levels of hormones. During fetal development those hormones play a relatively small tune. But at puberty, people with fully functional ovaries get a dose of estrogen as strong and focused as the solo note of a French horn. In response to that call, breasts and pubic hair appear, hips widen, and menstruation begins.

Streak gonads don't know how to play that note. At puberty their horns are silent. Because of that, in 45,X girls, secondary sexual changes don't occur, and only a very few of these people ever menstruate. Hormone supplementation can change most of that. And, using donor eggs, these people can have children.

In 46,XX people, both X chromosomes play parts in oogenesis (egg formation) and the development of the fetal ovaries. By the end of fetal development the ovaries contain as many as seven million oocytes. By puberty this number has dwindled to as few as four hundred thousand, and by menopause, fewer than ten thousand eggs survive. By contrast, in individuals with Turner syndrome, oogenesis begins normally in early embryonic development, but without a second X chromosome, millions of oocytes begin to die. And they continue to die through the rest of fetal development and for the first years of the child's life. By age two, there are no ova left, and the ovaries degenerate into fibrous streaks. In these women, menopause happens before menarche.[7]

In addition, people with Turner syndrome are usually short (under about four feet ten inches), have arched or "shield" chests, and webbed necks.[8] Sometimes the absence of that X chromosome affects people's brains as well, but rarely. Only about 5 percent of people with Turner syndrome have any degree of mental retardation.

Mosaics: Putting the Pieces Together

Over four thousand years ago, people began to create art using little pieces of broken stone. These mosaics usually contained stone quarried or culled from very different sources so that the end product had at least two dissimilar sets of fragments arranged together to create an image. Some of the greatest of these were found when the ruins of Pompeii were excavated.

Beyond the courtyards of Pompeii and the rotundas of our great houses of state, some people are mosaics too, and not in some literary sense. Through curious acts of living cells, some people have different numbers of chromosomes in different cells inside their bodies, like different pieces gathered from different stones. When those cells with differing numbers of chromosomes come from one zygote (one fertilized egg), this is called "mosaicism," and the affected people are living mosaics in a very real biological sense. When the mosaicism involves the sex chromosomes, we call these people sex-chromosome mosaics.

Sometimes, as the zygote begins to grow, nondisjunctions occur during the first few divisions. If that happens with the sex chromosomes, and one cell gets two X chromosomes and the other none, then as the individual develops, some of her cells are 45,X and others 46,XX.

Even though it seems that this sort of nondisjunction should create one cell line that contains three X chromosomes, this doesn't seem to happen. Other times, nondisjunction creates mosaics with three different sex-chromosome karyotypes, some even including normal male and female karyotypes, like 46,XY/45,X/46,XX or 45,X/46,XX/47,XXX cells—all inside one person.

All of these variants fall under the heading of Turner syndrome, which includes people with mosaic mixes of cells including 45,X cells with 46,XX cells or 46,XY cells. People who have cells with abnormal X chromosomes (usually X chromosomes that have lost some of their DNA) also fall under the heading of Turner syndrome.

During cell division, whole pieces of the long arm (called the q arm) or chunks of the short arm (called the p arm) of the X chromosome may get left behind, creating X chromosomes that don't do all that they normally do. These, coupled with one or more complete X chromosomes, can create karyotypes like 46,XXp- or 46,XXq- (missing pieces of the p arm or the q arm), 46,XXr (with a ring chromosome formed from a piece of the X chromosome), 46,XXqi (another abnormal form of the q arm), and many, many more. The people with these karyotypes can range from apparently normal females to females with hypoplastic (underdeveloped) ovaries to Turner syndrome females, to females with vestigial streak ovaries, to females who look like males, to males with dysgenic testes, to normal males.[9] Like light through a prism, as the X chromosome splinters, so do our old ideas about what makes a man or woman.

Chimeras: When There Are Two of Me

In Greek mythology, the Chimera was a fire-breathing monster with the head of a lion, the body of a goat, and the tail of a serpent. In literature, a chimera is something fanciful, often imaginary. And in science, a chimera is one creature made from the cells of two or more animals. The word itself is a chimera made from parts of many different histories. It is also a word that drives certain people's lives.

Mosaics develop from a single zygote. Chimeras, on the other hand, arise from two different zygotes.

Many mysterious things are possible during the earliest stages of embryonic development. For example, some eggs with two nuclei survive through to ovulation, and it is sometimes possible for two different sperm to fertilize these binucleate eggs. Occasionally, instead of splitting into two separate individuals, this sort of zygote develops into one person with two different types of cells—46,XX and 46,XY, for example—male and female all at once. On other occasions two separate zygotes may fuse and develop into a single fetus, again with two completely different types of cells—one from zygote A and one from zygote B, each with its own karyotype.

While these are the most common ways to create chimeras, they are not the only means by which a person can come to be a collection of two (or more) distinct types of cells, each with its own set of chromosomes.

The only difference between mosaic and chimeric people is that in mosaics the two cell types originate from a single cell and a single nucleus. In chimeras the two types of cells originate from separate cells of distinct genetic origins. Because of that, chimeras usually develop very much like their mosaic counterparts, creating a whole spectrum of people ranging from male to female.

Just how many of us are chimeric? No one knows. Often, sex chromosome chimeras show no symptoms at all. Everything depends on whether the chimerism extends to the gonads. If all of the cells of the gonads have two X chromosomes, the chimera develops as a normal female. If all of the cells of the gonads have one X and one Y chromosome, the chimera develops as a normal male, usually. However, if the cells of the gonads are mixed, then it is possible that the chimera might have one ovary and one testis, two ovotestes, one ovotestis and one ovary, or one ovotestis and one testis. And the ovotestis could be mostly ovary or mostly testis. The consequences of all this vary considerably. But, in the end, the external and internal genitalia can range from normal female to intersex to normal male and all the points in between.

KAILANA SIDRANDI ALANIZ
AGE: 38
LOCATION: WASHINGTON, USA

On October 10, 1970, the day she was born, she was named Dorothy Maree Alaniz—a baby girl. Curiously, though, no one filled out a birth certificate that day. When the certificate was finally filed on November 5, the name on it was Rudolph Andrew Alaniz. Within less than one month after her birth, this girl became a boy.

Then, when he was about eighteen months old, the doctors performed a laparoscopy on Rudy. Today, Kailana (born Dorothy/Rudy) feels certain that the doctors discovered then that Rudy in fact had ovaries (possibly rudimentary ovaries or ovotestes—gonads with both testicular and ovarian tissue) and that he/she was a 46,XY/45,XO mosaic. That meant that some of Rudy's cells had the normal number of chromosomes, forty-six, including normal X and Y sex chromosomes, and some did not.

Rudy's mosaicism raised some serious questions about his sex—was he a boy or a girl? His parents had always wanted a boy, and so all memories of Dorothy evaporated. Rudolph he would remain. So even though most Turner syndrome patients, especially those with any ovarian tissue, are raised as girls, Rudolph faced life as a boy.

According to Kailana, it seems probable that a few more early surgeries brought Rudy's body more in line with his parents' expectations, and the feminine pronoun disappeared from his medical records. Finally, everyone seemed sure of Rudy's masculinity—except maybe Rudy, who by age twelve was having serious doubts. His pediatrician and parents constantly asked Rudy if he was happy as a boy, and he repeatedly told them that he was not happy and felt like he should have been a girl.

But it wasn't until Rudy joined the army ten years later that he began to suspect how little he had truly been told about just who he was. Because of a back injury, the young man was sent for an MRI. That MRI changed his life completely. When the radiologists examined the MRI, they found an internal set of gonads, possibly undescended testes, possibly ovotestes—and an underdeveloped uterus. So the doctors had some blood drawn and ran a few tests. What they came back with surprised everyone. First, the blood tests showed that the young "man" suffered from congenital adrenal hyperplasia (CAH). The karyotype took a few weeks longer and confirmed that Rudy had two very different types of cells in his body; some were 46,XY, as they should have been, and others were 45,XO, as they should not have been. The final diagnosis was 46,XY/45,XO mosaicism and congenital adrenal hyperplasia—most likely caused by a congenital defect in an enzyme called 21-hydroxylase that leads to underproduction of cortisol.

Normally, as a fetus develops into a female it is exposed to only low levels of androgens from the developing adrenal glands and a lot of estrogen from the adrenal glands and the ovaries. But when 21-hydroxylase isn't functioning properly, low blood cortisol signals the pituitary gland to secrete more ACTH (the hormone that normally triggers cortisol production in the adrenal glands). That extra ACTH causes growth and division of the cells in the adrenal glands and overproduction of 17-hydroxy progesterone. That leads to overproduction of adrenal androgens (all produced by the adrenal cortex). During development, all of those extra androgens cause female fetuses to develop varying degrees of masculine genitalia. And, depending on many factors, in the presence of excess androgens, the uterus, the Fallopian tubes, and the vagina may develop differently.

After all of those tests, the doctors told Rudy that along with congenital adrenal hyperplasia and 46,XY/45,XO mosaicism, he also was a true hermaphrodite—he had gonadal tissues of both sexes. This diagnosis, complex enough for anyone, flipped everything inside out for this particular young person. That small change in one gene led to a world of new possibilities for Rudy, not to mention new fears and new anxieties.

In the end, both Dorothy and Rudy were laid to rest, and from those ambiguous ashes rose Kailana, a woman who identifies strongly with her feminine side, regardless of her past.

For reasons that aren't clear, Kailana has never suffered from hypercortisolism, or Cushing disease, a common affliction of people with adrenal hyperplasia. But that doesn't mean her afflictions haven't made her life difficult in myriad other ways.

I asked how her parents accepted her diagnosis.

[They] still will not acknowledge what I am or what they have allowed doctors to do to me [as a child]. They really just won't talk about it. The few times that they did offer me information, they denied [it] later. Yet, my extended family—aunts, uncles, and surviving grandparents—have told me a lot of what my parents won't, and [they] accept me. . . .

My parents . . . tell me I am making everything up, which has just made getting help from the medical community even harder. But it was them that told me my birth name of Dorothy Maree, they are the ones who said I looked like a girl when I was born, and my aunts and uncles say the same thing—everyone thought I was a girl, because that is what I looked like when born.

Also, living with a birth certificate that is dated a month after I was born has raised plenty of questions all by itself, for me and for doctors. All the medical tests in May of '93 just clarified why I have the things [and feelings] I do, why my blood says female, or kind of female.

A woman with androgen insensitivity syndrome told me she had experienced some of the same sorts of problems with her parents. There is something about sex and sex differences in children that lights a candle in a very dark room inside of many of us, a room we prefer to keep darkened. So when that flame flares, the easiest thing to do for some parents is to simply shut the door and pretend they never saw the things that glistened in that candlelight.

So Kailana keeps looking for hard data to try to pin down just who she really is. She notes, "It's like some hormone levels are in the normal female range [and] some just too high for [a] male. . . . I'm just in between the norms for male and female or in the norms for female."

Kailana sees herself as female, but she is also quick to acknowledge that she is a true hermaphrodite. Kailana moves between those worlds with poise and perception. Nonetheless, she still has some major doubts about what is actually going on inside her.

> Really, I am not completely sure of what is or was exactly done to me. What I find most annoying is the CAH diagnosis. Other than the right adrenalectomy, I am completely untreated [for CAH] and have been for the majority of my life. While at the same time, I am not exactly sure which of the "17" forms of CAH I actually have, just that it's the 17-OH progesterone that I over-produce, which makes for some really high cortisol levels when I'm under stress. What has me so perplexed is that I thought overproduction of cortisol was Cushing's syndrome. While at twenty-two years of age, I didn't show any signs of Cushing's, I was fairly fit, even with the screwy hormone levels.

Kailana does not take this lightly. She has learned a lot about congenital adrenal hyperplasia and pediatric as well as reproductive endocrinology. Still, getting regular and accurate information about her condition is a problem.

"I often wonder why 5 percent potassium chloride is added to my IV solutions, apparently a normal IV solution causes my body to go into ketonuric shock or something like that. I don't even know if that is the right term."

Overproduction of cortisol interferes with normal glucose metabolism. Normally we derive the majority of our energy needs from blood glucose. If high levels of cortisol prevent that, then we switch to digesting fats. Digestion of fats produces a group of compounds including acetoacetate, beta-hydroxybutyrate, and acetone. Collectively these are known as ketone bodies, and the acidic character of some of these can cause the blood pH to drop dramatically—the blood quickly becomes too acidic to support life, and the patient goes into shock. As the process proceeds, large amounts of ketone bodies appear in the urine—ketonuria—and the patient is said to be ketonuric.

> What everyone doesn't get is that doctors don't tell me anything, and I find out afterward that they have a plan to use

different meds if I go into seizures or if I go into the ketonuric shock; you see, my medical records state things that [the doctors] won't actually talk to me about. I know I have to worry about things, and I know it has to do with my one adrenal gland, I assume it's to do with CAH, but I really don't know if it is, in fact, CAH. The military doctors were, [and still] are the only doctors to acknowledge what I have, what I am. The rest, well they're not much help. And I really worry sometimes, that as they see me as male, they aren't particularly worried about the CAH condition, I'm not on a treatment plan. So at thirty-six, I am showing signs of things that are kind of common to both CAH and Cushing's, yet none of them will do anything to help me explain which it actually is.

I know my right adrenal gland was removed as an infant, and [I] suspect some penile reconstruction with other genital cosmetic alterations. But I really don't know the exact dates or why [these things were done] other than being told my sex was changed from female (what I looked like when born), to male (how I was assigned on [my] birth certificate) a month later. I would imagine that those surgeries were much later. I do have some really odd illnesses that kept me extremely underweight as a toddler.

I have MRI scans that show how I am formed, how the phallus is shaped, and it is not normal for a male. . . . I have spent a great deal of my life being teased about how I am developed. Lots of jokes; it has pretty much left me with a low acceptance of myself as a male. . . . I have had a life of questioning my gender—gender identity disorder [GID]. As a young kid, [from] ten years of age on, I have known I should have been a girl. Constantly being questioned by my mother and family pediatrician about being happy as a boy, and me telling them over and over that I am one of the most miserable people on this planet. [That] I should have been a girl made absolutely no difference to them. They just ignored me.

I am almost thirty-seven, I am still single, I have never married, and I do not date. I am a little over fourteen years celibate, which is probably my biggest complaint. I am terribly ugly looking. I am extremely [camera shy], I don't look even close to having a female appearance. Yes, my breasts are developing [she is now taking estrogen supplements], but a great deal slower than I had expected. My body hair is a lot lighter as well. Other than that, though, I have way too much muscle, lots of body fat, I'm slightly obese. My neck is as thick as what you would expect on a three-hundred-pound linebacker, my chest, extremely broad.

My family relationship is extremely strained. I don't go to family events—Thanksgiving, Christmas, birthday parties—nor do I celebrate my own birthday. I feel as though I am the black sheep of the family. . . . Losing family over the way you're born isn't right, but the philosophy doctors keep preaching on how we should be raised, treated, have our medical records withheld from us. The lack of information and acknowledgement of what we are has made me a very bitter person, especially when dealing with doctors.

In a way, being intersexed, true hermaphrodite, has given me a reason to transition with confidence . . . that this is what I should have been [a woman] all along. I guess having a genetic disorder that says you're part female—anatomically part female—makes transitioning a lot easier from a personal-acceptance point of view. It's like there is more than just me believing I am a woman. That provides comfort to me. . . . I really think that everyone knows what they are, and can comprehend just how confusing it would be to question what they are, knowing they had surgeries, knowing they had extra anatomical things, or genetics, and finally discovering as an adult that they are different from most people.

While no one seems to be willing to acknowledge what genetic condition I have, what was acknowledged during a Social Security income review in September 2007 is that I have a genetic disorder—a "multi-body system disorder" that causes delusional gender identity issues along with the depressive states. . . . That information came from medical records that were read into that hearing, reviewed by an orthopedic doctor, questioned by a psychiatrist, and witnessed by a federal judge.

As for being delusional or GID, I don't know. But at least what was said shows that I actually don't fit that diagnosis well. Still, the genetic condition, the amount of trauma and depression, and eccentric behavior patterns, all make it difficult to fit socially. . . . So far, that is the most information, on record, that I have ever been able to get acknowledged about my genetics and anatomy from anyone since returning home after my military service.

I wish the final findings of fact had actually stated that I was, in fact, intersexed, or a true hermaphrodite. Instead a genetic condition was used, and a multi-body system was used, as terms to explain what I am. . . .

Truthfully, I think the most important thing I would like people to understand about me is that I am a person, I have a right to my own body. It is mine after all, I am the one that has to live with it, and no one else has a right to make decisions for me. I really hope people come to understand that each of us has that same right. There isn't a person on this planet who should be forced to live as something they aren't. While doctors still continue to spread the belief that assigning a child as a girl or boy is extremely crucial to their well-being, for those of us who they chose [the] wrong [sex for], our lives are just tortured. And for [those of us for whom the doctors] chose correctly, there are still huge emotional conflicts and emotional issues that are continually being ignored because of the medical standards in practice for intersexed persons.

We are placed into a world that has the openness and understanding to accept an intersexed person as a perfectly normal human being, and it is the fault of the medical community that has shunned, shamed, hidden, tortured, humiliated, traumatized, and continually discriminated against us because of how we were born, to the point that there are those people who remain ignorant about people born with ambiguous genitalia, chromosome disorders, and hormonal variations that make us the unique people we are.

Bits of Y Chromosomes: Lost and Found

Like with X chromosomes, Y chromosomes may also lose parts of themselves, duplicate parts of themselves, and pick up stray pieces of other chromosomes that they find lying about in the nucleus.

Y chromosomes are an unusual assortment of treasures and junk. The chromosome itself—at least in idealized textbook illustrations—looks a little like a bowling pin. We call the short upper part the short arm and the long, fatter, lower part the long arm of the Y chromosome.

About half of the long arm of the Y chromosome seems to be junk and contains no genes at all. The short arm, on the other hand, contains some genes that can make all the difference between men and women. This part we named the sex-determining region of the Y chromosome (SRY), and it contains genes that will, almost all by themselves, turn an uncommitted pair of gonads into testes and a fetus into a boy.

Normally, SRY shows up exclusively on an intact Y chromosome, but not always. Occasionally, during development of sperm, a Y chromosome will drop a piece of itself on the cutting-room floor. If that piece happens to include SRY, then this particular Y chromosome has lost its knack for making baby boys. When that sperm hooks up with a normal ovum, the resulting zygote, embryo, fetus, and child will have the 46,XY karyotype, but, in every other way that matters, it will be a girl, or almost. This change comes in two forms called complete and partial gonadal dysgenesis (dysgenesis means abnormal development). 46,XY people with complete gonadal dysgenesis have fully formed female external genitalia, but the ovaries are not completely developed and are usually present as streak ovaries. People with partial gonadal dysgenesis have incompletely formed testes and varying degrees of partly formed male external genitalia. All of it seems to depend on just how much of the Y chromosome gets lost.

A similar situation can arise when a mutation destroys portions of the SRY. These individuals, also 46,XY, also develop into baby girls. One example of this sort of change results in complete gonadal dysgenesis. A gene known as "desert hedgehog" lies in the SRY region of the Y chromosomes. Desert hedgehog is a member of a gene family that includes sonic hedgehog and Indian hedgehog—all of which play key roles during early pattern formation in the developing embryo. When desert hedgehog fails to do its job, 46,XY fetuses develop normal vaginas and uteruses but only streak gonads. At puberty, these people do not develop female secondary sex characteristics.[10] Clearly the genes within SRY play major roles in the development of male babies.

Interestingly, though, only about 15 percent of 46,XY females appear to arise from loss of or mutations in the SRY. So SRY is only part of the story—a necessary but not sufficient piece along the twisted path to boyhood.

On occasion, a man's X chromosome will find that piece of Y chromosome left behind in the nucleus, then pick it up and stick it in an empty pocket. That bit of greediness creates an X chromosome carrying the SRY.

When a sperm with that X chromosome hooks up with an ovum carrying the usual X chromosome, then the zygote, embryo, fetus, and child will have a 46,XX karyotype. But it will enter this world as a baby boy, or almost. Even though everything looks boyish at birth, as adults, these people are sterile and generally have smaller than normal testes.[11] Otherwise, they are pretty much just like 46,XY people.

True Hermaphrodites

To zoologists and botanists, true hermaphrodites are animals or plants that can play the role of either male or female, sometimes even both, during sexual reproduction. The ones we are most familiar with, like hermaphroditic trees, can even fertilize their own ova. By those terms, there is clearly no such thing as a true human hermaphrodite.

Regardless, there is a group of human beings that reproductive biologists call true hermaphrodites. By definition, a true human hermaphrodite must have both ovarian and testicular tissues either in the same or different gonads.

Because of the inaccuracy of the word *hermaphrodite* in reference to human beings, scientists have recently suggested that we refer to these people as persons with ovotesticular disorders of sexual development. Because of its historical significance, however, in this chapter I will stick with *hermaphrodite*.

The most common karyotype among true hermaphrodites is 46,XX (one of these X chromosomes usually has the SRY it collected from a Y chromosome) followed by 46,XX/46XY chimeras or mosaics, and only rarely 46,XY.

On the outside, these people may look like males or females or something in between. Almost all of them, though, are raised as males because they have enlarged phalluses. Those phalluses—even if they began as clitorises—grow to nearly penis proportions because of the testosterone made by testicular tissue, always a part of a "true" hermaphrodite. Vaginas and uteruses are present in the majority of these people, and usually at least one testis is palpable in the labioscrotal fold (that is, one testis has descended into what is a mixture of scrotum and labium). Breast development occurs with most true hermaphrodites at puberty, as does menstruation because of the ovarian tissue. But very few true hermaphrodites produce sperm. Ovulation, though, is common, and many true hermaphrodites with 46,XY karyotype are fertile and have borne children.

For most people the cause of their hermaphroditism never becomes clear. Several possibilities exist: sex chromosome chimerism and mosaicism are among the leading possibilities. In either of these cases, one of the gonads could be 46,XX and the other 46,XY; or individual gonads could contain cells with both karyotypes. Alternatively, if the SRY region of the Y chromosome was picked up by another sex chromosome or even a chromosome other than X or Y, SRY could create a similar situation with cells of two different karyotypes appearing in one or both gonads.

Regardless, by their position in the middle, true hermaphrodites reveal to us the full array of the human spectrum lying in between the unattainable ideals of the perfect male and the perfect female, between XX and XY.

Lisa May Stevens
Age: 42
Location: Idaho, USA

"The nasty, nasty crows are so very cruel to her, just because she is so very different . . ." —from one of Lisa May's poems

"Mother was a little strange at best," says Lisa May Stevens. "When I was between five and six years old . . . she would call me to her room, strip me naked, and begin to dress me in my 'Lisa May' clothes. If I objected, she would pull my dress up and spank me."

At the time, Lisa May's name was actually Michael. Lisa May was his grandmother's name, but at age five, Michael had no idea why his mother called him that. At first, Michael resisted these overtures from his mother. In return, he was spanked, jerked around the room, and tossed into a chair where he could only sit and cry until he agreed to his mother's requests. Michael's mother always outlasted him. Her insistence on treating Michael like a girl slowly worked its way inside of Michael's head. He began to like the colors she picked for him, the panties and the petticoats, the dresses and the bows.

"If I could please her . . . she, for a while, would be happy. And she was so much fun to be with if she was happy. If I pleased her she would primp me. And on rare occasions she would put makeup on me, but never too much . . . otherwise you might get mistaken for a slut."

Michael, of course, played along. It was the only way he could hold his mother's attention, and he craved that attention. As Lisa May, he was loved. As Michael, it was less clear.

> I never knew what I was for most of my life. My father treated me like I was all boy. He could be soft, but mostly he taught me to be tough and strong inside and out. I wasn't allowed to cry

or throw tantrums in front of him. Repercussions were severe—from either one of them. If I acted like I did with mother in front of my father, I would get it. And it was the same the other way around—she was so feminine all the time, and father was so masculine. It was a hard balancing act being around the both of them at the same time. So, I spent as much time as I could outside hiding or, if possible, at a friend's house.

Years later, when I visited with my childhood friends and parents, it seemed they all knew what my mother did with me—dressing me up as a girl. They wondered how it would affect me later in life. They all thought I'd grow up and be a cross gender. Well, they weren't far off. I grew up and found out I was intersexed instead.

Like so many intersex children, Michael was kept in the dark. He had no idea that he was physically anything other than a boy. It would take years, and a major mistake on the part of an attorney, before Michael would learn the whole truth.

Michael's mother had had several miscarriages. So while she was pregnant with Michael, she took a drug called diethylstilbestrol, or DES. From the 1940s until the late 1960s, doctors routinely prescribed DES for complications and maintenance of pregnancy.[12] Some two million women in the United States and maybe as many as four million women worldwide took the drug while pregnant. As early as 1953, studies at the University of Chicago suggested the drug had no effect on pregnancy. But doctors continued to prescribe DES until the early 1970s, when it became clear the DES was carcinogenic and closely linked to the development of clear-cell carcinomas of the vagina, endometrial cancers, and breast cancer.

In the late 1980s, lawyers filed a class-action lawsuit on behalf of those damaged by DES. With all the coverage in the news, Michael's mother became concerned and interested in the lawsuit. She contacted the lawyers involved with the case, and when she mentioned that her son was somewhat out of the ordinary, the lawyers expressed a great deal of interest. They were especially interested in children with cancer.

Michael had had cancer—twice, in fact. The first tumor appeared when he was seventeen, a gonadoblastoma—one of his undescended testes had turned malignant. The doctors performed surgery after surgery, ten in all, before they finally gave up and removed the testis. That seemed to take care of things. But shortly after Michael turned twenty-one, he began to lose weight fast. In a few months he dropped from 180 to 125 pounds. At first the doctors could find nothing wrong. As a last resort, they tried a needle biopsy of his remaining testis and its surroundings. They found a seminoma—a tumor of the gametes.

Michael immediately went home and told his wife, who shed a few tears, went into her room for about fifteen minutes, then came out with her suitcases packed and said "Sorry, dear, I can't sit here and watch you die." She left, and a short while later filed for divorce.

"I went crazy. . . . I took seventy-five thousand dollars out of my savings and went all over the world for three months wishing to just end my torture. I could not at that time accept being a eunuch and not knowing what was to come of me."

Finally, Michael wore himself out and landed at a friend's house. While he was there, he met a woman who gave him the strength to go through with the surgery and the chemotherapy. That surgery would rewrite Michael's history and his future.

During the surgery for his cancer, the doctors removed what they thought was Michael's remaining testicle. Instead, they found an ovo-testis. The implications were dramatic. They offered Michael free karyotyping so they could further investigate the true nature of this man.

They took a few cells and sent them off to the lab. Simple enough, but inside of those few cells Michael's lie was hiding.

The doctors could barely contain their excitement. They asked Michael to go to a medical center at UCLA for further testing. He agreed, and spent eight days undergoing all sorts of tests. And then he waited.

When the results finally came back from the lab, Michael was no longer a man, not yet a woman, and more than he had ever imagined.

Michael, it turned out, was a true hermaphrodite—a 46,XY/46,XX chimera. At birth, he had both ovarian and testicular tissue, a penis, a uterus, and labioscrotal folds that the doctors had cut and stitched into something like a scrotum. He had become a chimera, literally, when two individuals fused into a single being, the result of what doctors call a "tetragametic fusion" event—meaning that when Michael's mother became pregnant, she carried two fertilized eggs inside of her that would have become fraternal twins—one boy and one girl. Instead, those twins joined and became a single being. Michael even had two blood types—B-positive and O-positive.

That day, inside Michael, Lisa May's heart began to beat a little more strongly.

> This was about the time Mother got a hare-brained idea that the DES was the cause of my cancer. She convinced me to join a class action. And that's when the lawyers got all of my records from her, including all the medical records that described all the things done to me up to that point.

The lawyers quickly realized that Michael's true hermaphroditism (a term Lisa May prefers to intersex) and chimerism eliminated him from any consideration for reimbursement by the courts. At that point the attorneys turned all of Michael's records over to Michael.

For the first time, Michael saw that not only was he not like the other males he knew, but he was the last of many to know this. Inside his medical history, he saw for the first time that his parents and doctors—not his chromosomes, not his hormones, not his genitalia—had made him a him. For the first time he understood the web of lies his parents and doctors had spun around him.

At his birth, everyone saw the ambiguity. There had been arguments, discussions, and surgeries, and where there had been doubt, the doctors had sewn the seeds of certainty. A few stitches to pull the labioscrotal folds together into a sort of scrotum, covering the rudimentary vagina and uterus, and some work on the urethra, and the deal was done. Michael's parents' silence sealed the few remaining cracks, and Michael became a boy, to them at least.

At the time he made these discoveries, Michael was a practicing Catholic. He explained to his priest what he had found and how it made him question himself and God, and even entertain thoughts of suicide. His priest responded that even though the Lord generally condemned those who committed suicide, in this instance He might make an exception.

All but two of Michael's friends dropped him. Suddenly he was alone, scared, and uncertain about even the most basic parts of himself. But somewhere beneath all of that, Lisa May was opening her eyes, taking long deep breaths.

Everything changed. What had once been solid ground shimmered with quicksand.

> It was after I . . . found out what I was that I began feeling sorry for myself and tried to bury everything in my work, that I confronted my mother. She told me everything she could or was able to, but she was at best hostile to me. . . . But she did remind me she had taught Lisa May everything she knew, and if I desired I could always be Lisa May since I had no male parts left to speak of. At the time I, of course, got offended at her. But over the next few weeks, I got some makeup and girl clothes and started to practice being Lisa May again. And, wow, to my surprise, everyone really liked her, and I made a new bunch of friends. Mostly they were TSTGs [transsexuals and transgenders] and lesbians. They didn't seem to care what I was. That made me happy. So at twenty-two years old, I started to become Lisa May again.

Lisa May's rebirth happened so quickly that some aspects of womanhood caught her off guard. One night at a party with some of her newfound friends, a drunk man came on to her. Lisa May didn't understand that she was being propositioned, and her discomfort was mistaken for rejection and ridicule. The man shot her in the leg with a twenty-two-caliber pistol. As it turned out, that was one of the more minor events her intersex would trigger.

A few months later I was at another party, and a different guy hit on me. But this one really got mad. By then I belonged to a dyke. She was my lover and companion. Melissa [not her real name] was my whole world. . . . . After a while, he came back and walked up behind me. He called me a bitch and a whore, and as I turned to see what was happening, my lover tried to shove me out of the way, but he fired his .38 snub-nose revolver at me. The bullet hit me just under the left eye. . . . I had never seen so much blood. . . . I went through four plastic surgeries and had the very best doctors. . . . They made me even better than before. [When I found out later] I was amazed that Melissa tried to take the bullet for me. . . .

So little Lisa May's coming out wasn't so great at first. Lack of training on my mother's part, I guess. Nine months later I was healed up from all the surgeries, and my lover became obsessed with protecting me. . . . I was so in love with her, she treated me like a princess. . . . I took care of the home and the small yard. I grew fresh vegetables and herbs so I could cook fresh stuff for Melissa whenever possible.

Lisa May and Melissa spent as much time together as they could. But their happiness would not last long.

[Melissa] and some TG friends of ours decided to go to Ensenada, Mexico, in our friends' RV. I was so happy; it just couldn't get any better. We went horseback riding, I snorkeled a little, and got sunburned, but it was like a honeymoon for us. On the third night, Melissa and the others got really drunk and ran out of booze. So little Lisa May offered to drive to town and get some more liquor. [Lisa May does not drink.] It was really late, but someone said [the stores] would most likely be open. I was dressed in a really nice yellow sundress as it was warm there. . . . So I got the car keys and drove to town. I didn't realize it was so late. When I got [to Ensenada], the town was closed up, nothing was open. So, because I had said I'd bring the booze back, I thought, well Tijuana wasn't that far off, and they're open all night. So off I went, singing to the music and happy as I could get.

I was driving for about twenty or thirty minutes when the car started to sputter and then quit. I pulled over to the side of the highway . . . got out of the car and popped the hood and was leaning over the engine when a truck pulled up and stopped. . . . I thought they were the Federales. I waved them I was fine, and they drove down the road. I went back to looking under the hood of the car and suddenly I felt a horrible pain shoot into my head and down my body. I didn't pass out, but I fell to the ground, and I heard several voices in Spanish yelling and whooping it up as they kicked me without mercy everywhere. After what seemed like forever they tore my dress into pieces. I tried to hold onto the pieces of material but they were still kicking me and such.

At this point Lisa May's assailants realized that Lisa May was not exactly the girl they had expected her to be. That incensed them. They threw her into the back of the truck and drove off.

They stopped finally, and I knew I was going to die then and be buried out there somewhere. They pulled me out of the truck and took me and took me and took me all night long. I don't remember when I passed out for the last time, but I do remember they beat me ruthlessly all that night. . . . At some point with my face so swollen I couldn't see any longer, it seemed to be getting light out. I heard, I think, their truck doors open and close and their motor start up. Then, one of them walked up behind me, I think, and pulled my head back. I opened my eye then and saw a knife going in front of my face, and he slit my throat, and threw my body on the ground.

I felt as if I were floating in the air. I think I woke up several times that day, but I could not move. . . . I felt my well of life draining out of me. I prayed to God to end this torture and take me home. . . . A while after it got light, I heard voices saying here she is, and a girl came up to me and turned me over. I opened my eye and thought the angels finally decided to bring me home. She said "be still" and tried to give me water. . . . I don't know or remember how I got to the San Diego hospital, but that's where I woke up, with IVs in me, and tied down. I really tried to freak out then, but I couldn't move my body.

Lisa May spent eleven days in the hospital. The doctors wanted her to stay longer, but she couldn't stand another day. She wanted little to do with Melissa just then, and that, coupled with the horrible things Lisa May had suffered, broke Melissa's heart.

For a month Lisa May kicked around depressed, blaming herself for what had happened to her, hating God for having made her what she was. And then Lisa May tried to kill herself with pills. All that got her was a week on the psych ward.

After the rape, Lisa May couldn't stand for anyone to touch her. No one was ever accused of the crime. No one besides Lisa May and her friends knew what had happened to her that night. Lisa May continued to spiral deeper into depression and stopped speaking. Several weeks later she checked into a Buddhist retreat for a three-month stay. Melissa visited her there, but still Lisa May would not speak.

I just sat there, staring off into heaven. [Melissa] had gotten so depressed . . . no matter how hard she tried she couldn't wake little Lisa May up. . . . One day a staff person came to me and talked to me, but I didn't answer until she told me my Melissa committed suicide the night before. . . . I was destroyed after that. I left the retreat and went camping for a few days and prayed and said my good-byes to this world.

At the end of the third day, I loaded up my pistol with a special round. While watching the sunset, I put the pistol up

to my throat under my jaw and pulled the trigger. . . . I heard
it just click as the hammer came down on the shell. . . . I was
mad as hell. I couldn't do even that right. . . .

   After a couple of days of swearing at God . . . I left the
mountains, moved in with some friends, and saw a doctor and
got a lot of testosterone and started to inject myself and got big
and strong and went to work as an ironworker for some time
after that.

Lisa May continued to inject massive doses of testosterone. Her muscles gained new bulk. She took martial arts lessons. At first she focused
solely on revenge—a hard, slow steely revenge against those who had
raped and beaten her. Eventually, she came to see the folly in that, but
still she worked to bring Michael back into full control over her tortured
body and mind. During this time Michael married three times—each
time to a lesbian woman—and each time the marriage failed.

Ever since Melissa, I have had rotten luck in relationships. No
matter how hard I try to please my mate, it only takes one bad
comment about my body and I start to shut down my emotions. If the negative words or gestures continue, I just retreat
into myself and hide in a place in my mind that is a safe place.
. . . I feel I'm ugly as hell ever since that nightmare in Mexico.
I feel nobody will ever love or adore me as Melissa did, and I
feel I can't please anyone anyhow. . . . [The rape] really screwed
me up as far as being touched by anyone in a loving way, even
though I knew it was in a loving way, and not in a forceful and
hurtful way.

Only in the last couple of years has Lisa May's life begun to change
significantly. For years she has wrestled with the fact that she does not
look or feel like a lot of other people, wrestled with her differences and
her shame and her fear and her loneliness. Time and again, she has
drawn from a very deep well to find the courage and the bravery to go
on. She is a remarkable woman.

In the end, as she puts it, Michael let Lisa May out. She stopped taking the testosterone and started thinking about estrogen instead.

The decision to become and live as Lisa May full time is not a current one but one from a lifetime of conflict. Since I have no male side really left [in her mind, at least], and I'm growing breasts and such, and at least half or more of my body looks female, and I move, walk, gesture, and such, and I thought, why not?

It seems I don't fit into the male role very well anymore. So this way I can at least be something instead of an "it."

The hormones are really kicking in, and . . . soon enough, I will have my implants. So after that, I am Lisa May all the time. And I'll be moving, so no one will know of me anyway. So my journey to find some peace of mind and happiness is almost upon me.

The demons still come for Lisa May, especially late at night, and especially when memories of Mexico and Melissa flow back to her. But the demons are losing more often now. Lisa May is getting stronger. She's also told me that her sexual experiences are unlike any of those described to her by men or women. Hers are, she says, a much more intricate mix of male and female responses. Hers are much better, she says. She told me, too, that she just won a "best legs" contest on one of her first nights out in a very long time.

"I am proud," she says at last, "to be a hermaphrodite."

Pseudohermaphrodites: Beyond X and Y

Pseudohermaphrodites are very different from hermaphrodites, who have gonadal tissues of both sexes. Here, *pseudo*, meaning "false," refers to the external appearance of these people, which does not correlate with their karyotype. Female pseudohermaphrodites have a 46,XX karyotype and have ovaries, but their external genitalia include fully or partly formed penises and scrotums. Male pseudohermaphrodites are

46,XY, but their external genitalia are partly or fully female. Male and female pseudohermaphrodites usually have normal sex chromosomes but carry abnormal genes on other chromosomes.

DIANNE
AGE: 39
LOCATION: ONTARIO, CANADA

Dianne was born to an unwed mother at a charity hospital in 1969. "There is reason to believe that there may have been some degree of physical abnormality," she says, "but no records exist (or are admitted to exist) about my earliest life. About six months later, I was put up for adoption as a boy. Boys were much more in demand for adoption in those days."

Though advertised as a boy, as far back as she can remember, Dianne says, she was certain she was a girl. "I spent most of the first decade of my life with absolute assurance that I was a girl," she recalls. "I played with other girls exclusively and disliked the rough and rude behavior of boys. In the beginning I was told I had a particular affinity for flowers and for singing and could often be found sitting in the flowerbed and singing to myself. The earliest picture of my childhood shows me in a bright yellow dress with a profusion of blonde curly locks, and a huge smile." Dianne's earliest memories "are of tea parties with the girl next door and of being led, hand-in-hand, by the girl across the road who took on the role of my big sister."

Dianne grew up at the edge of a small town in Canada. At first, no one, not even the couple who adopted Dianne [she did not disclose her early boy's name] seemed worried about the little boy's girlish behavior. He was different. Folks recognized that and most accepted it—that is, until he reached school age. Then people's attitudes began to shift.

> Somewhere around the time I was to start school, my behavior
> became a cause for concern, and my adopted mother started to
> push me to "be like other boys." I found that so terribly confus-

ing. She may as well have asked me to "be like a fish," because I had absolutely no idea how to be anything other than who and what I was. And I still thought I was a girl. Sure, I had an "outtie" [an unusually large clitoris], not an "innie" [a fully developed vagina], but I just thought some girls probably were that way and that everything would be set right at puberty.

At school, Dianne played mostly with the girls and avoided the boys. But after a year or two, the girls lost interest in Dianne, who according to the school records and rolls was a boy. But Dianne had no interest in the boys and soon fell in with the few other outcasts who lingered in the empty spaces on the playground. And she began to understand that she wasn't exactly like the other kids; something about her "was wrong." It would have been a frightening discovery for anyone; it was a horror for an eight-year-old child.

While playing with a male cousin on the farm, he said to me, "You should have been a girl." I said "I am!" and he said "No you're not, not really." . . . I started to feel like I was just weird, somehow deformed. The pressure from my adopted mother to "behave properly" (like a boy) was also increasing, a course she tried to force using emotional abuse and badgering ("You are defective. I should have taken you back to the adoption agency."). I was always very open about what I felt, and the conflicts with my mother grew more heated over the years.

Dianne's adopted father was more understanding. He often found excuses to take Dianne away from home, away from her mother. The times they spent together were different, there was compassion and understanding, there was gentleness. But Dianne's adopted father would not or could not stand up for Dianne in the presence of her adopted mother. Only her grandmother ever found the nerve to face up to Dianne's adopted mother. A lot of words flew back and forth between the two women, but in the end Dianne's mother always had the last word. And that word was *boy*.

Dianne knew otherwise.

At about age eleven, I had my first sexual experience with a boy, a relative a few years older than me who was attracted to my femininity, and, shortly after, a sexual relationship with another boy (also a relative). I've never revealed this before, to anyone, because I knew I wasn't gay, and everyone would assume I was if they knew.

I wouldn't consider what happened to be "sexual abuse" because we were little more than children, just old enough that we were developing sexual awareness and experimenting with our own sexuality. My own "experimenting" also included girls, which was pleasurable but not as all-consuming as being with a boy.

At about age twelve, a lot of things started to happen. Our home was annexed into the nearby town, and we had to change schools at Christmas time. This turned out to be a Godsend! I had been very quiet and terribly shy, but with my new group of friends I started to come out of my shell. I found a greater acceptance among the new girls. The boys were also more accepting of my femininity and treated me more as one of the girls. I also began to develop a deep friendship with one of the new boys, a friendship that turned into my first crush and then into first love!

It was also about this time that I experienced the onset of puberty. My faith had always remained that at puberty everything would be straightened out, that I would become a girl.

Unfortunately things didn't happen that way. About age thirteen I started to develop breasts, which pleased me to no end. But I also began to grow fuzz on my chin, which was absolutely devastating! My "outtie" didn't become an "innie," and that was the most distressing of all! I knew something was *really* wrong. While the boys were becoming men and the girls were becoming women, I had little boobs and an "outtie" that didn't even develop like the boys'. My "boy bits" had always been different, *very* small, and didn't change at puberty, nor did I develop normal hair growth. Aside from a luxurious head of hair, I had only a few fine hairs on my face and a light pubic fuzz.

At home Dianne continued to keep a low profile, and her mother continued to act as though nothing was happening even as Dianne's bed piled over with stuffed animals, her dresser gathered makeup, and her closet filled with dresses. But when she was away from her mother, Dianne began to dress more and more like her girlfriends. She wore makeup; her behavior became more and more feminine. For a while, that worked. But though she was playing it all close to the vest, Dianne was on a collision course with herself.

It was also around age thirteen or fourteen something else of critical importance happened in my life. While spending the evening with the boy I had a serious crush on, we were laying on the floor listening to music, he took me in his arms and kissed me—not a tentative "peck" but a full-blown passionate kiss, full on the lips! The feeling was like being struck by lightning! All of a sudden I felt passion, sexual energy like never before. I knew that what I was feeling were "girl feelings," not "boy feelings," and, if it hadn't been for my physical impediment, I would have lost my virginity that night. . . . My "crush" had suddenly turned to "first love." After the kiss, my boyfriend released me, stared into the carpet, and said "Please change into [something less alluring] before . . ." I went into the bedroom to change and had a good cry.

Dianne continued to love that young man for the next ten years. But every one of those years was filled with frustration as well as desire. Dianne's "condition" stood between them and any traditional consummation of that love. But the kiss and that love convinced her; she wanted her boyfriend, their intimacy, and the possibility of marriage and children.

The forces tearing at Dianne overwhelmed her, but when she asked her mother if she could see a doctor about her problems, Dianne's mother refused. There was, she felt, no real reason Dianne needed to see a doctor. For over a year, Dianne continued to plead her case, until she wore her mother down and was allowed to see a doctor.

It turned out that Dianne's testosterone and estrogen levels were both low. The physician suggested to her mother that Dianne start taking a hefty dose of testosterone to "make a man out of her." Dianne flipped.

"I'd rather take cyanide!" she told her mother.

That was 1965. By then at least some doctors knew about cases of unusual sexual development. But this particular doctor didn't pursue the matter of Dianne's sex. It might have been partial gonadal dysgenesis. It could have been bits of chromosomes slipping from place to place, or it might have been any number of other things, some of which we have names for and some of which we don't. At the time, no one seemed to care enough to follow up with Dianne. Regardless of the cause, the solution seemed obvious. As it turned out, Dianne never did uncover just how she differed from other boys and girls. For her, too, the exact cause became irrelevant. Resolution was the real issue. She refused the testosterone.

Since age five, Dianne had been regularly running away from home, usually to an abandoned cabin she had discovered in nearby woods. There she would live on vegetables she had stolen from neighbors' root cellars. After three or four days—when she figured her mother had calmed enough to allow her back into the house—Dianne would return home.

That pattern began to change when Dianne turned fifteen. One day, while listening to a radio program, Dianne heard the story of a person classified as a female, but who felt certain that she was a he. Suddenly Dianne no longer felt alone. Someone else had a problem just like hers, only in reverse. She wrote to the radio station, and the people there forwarded her letter to the person Dianne had listened to. He wrote back, and Dianne ran off to Toronto to meet him. In Toronto there were others, lots of others, some just like Dianne. And some of these people's parents completely accepted that their children were different. Soon Dianne had a new place to run to. Now she ran to the homes of understanding friends and understanding parents. Not only that, but in Toronto, Dianne could live as a girl. No one there had ever known her as anything else. She began to go every weekend, changing her clothes on the train as she sped toward her friends.

It was a delight to be able to meet other teens, go out to clubs and dances, and just be a normal girl in every way. Going home was always hard. Changing on the train into clothes I didn't want to wear, trying to wash off the makeup, trying to arrange my hair in a less feminine fashion. It always felt like going back to prison.

At about age seventeen, I had finished high school and spent more time in Toronto, where I met a young man [Jerry] who was quite smitten with me. He was from a well-to-do family of business owners, and we started moving with a different crowd, the young "up-and-comers." Instead of clubs, it was dinner parties, long dresses, and cocktails. For a country girl, this was really heady stuff! I was really beginning to develop as a person, as a young woman. I was no longer shy but outgoing, funny, and mischievous. Looking at pictures from that time, I also realize I was very pretty.

I didn't discourage Jerry's attention, so the relationship continued for some time. At one particular party, when the other couples had begun disappearing into private rooms, Jerry and I found an empty room and lay on the bed talking and necking. I had always set strict limits on what I would tolerate for intimacy—obviously—and Jerry had always accepted that. But after some heavy petting and encountering my limits again, he asked me to marry him. I said no, that I wasn't ready to settle down. He said we didn't have to marry right away but that he wanted me to say yes. I wouldn't, and Jerry kept pressing for a reason until finally I told him that I *couldn't* marry him and why.

He listened to what I had to say and said he *still* wanted to marry me! I explained that I was underage and that my mother wouldn't allow any medical treatment and would undoubtedly have him thrown in jail if anything happened between us. I told him that I had heard of a doctor overseas (Morocco) who would perform surgery but the cost was astronomical. He said that he would pay all the expenses if I would only marry him. I reminded him that I was a minor. He said that didn't matter—we'd move to another country. DAMN! Here was a chance to escape my deformity, but I would have to sell my soul to take advantage of it! I declined again.

But Dianne did accept Jerry's offer to pay for a trip to New York City to meet with a doctor Jerry had heard of who seemed to know something about people like Dianne. In 1967 she met with this physician, the first who didn't think she was nuts. He did a complete physical exam and ordered several blood tests. Apparently he thought Dianne had got it right, that she was a girl, because he wrote a prescription for estrogen. But Dianne couldn't get that prescription filled in Canada, and she couldn't imagine that any Canadian doctor would duplicate the prescription for her, nor did she want to risk trying to smuggle the estrogen across the border. So, while she felt the doctor's verdict had vindicated her, at least partly, she was still without her hormones, and her relationship with Jerry was beginning to close in on her.

"I liked Jerry a lot," she said, "and the lifestyle he offered was so very tempting. But I was not in love with him, and I was enjoying being a vivacious and independent young woman for the first time, free of shackles and limitations. But Jerry was very much in love with me, and it pained me to be a tease or a temptation. So I eventually slipped away home."

Because Dianne's birth certificate said she was a boy, she was accepted into technical school as a boy. But when she arrived, everyone assumed that she was a girl pretending to be a boy in order to get into technical school. She could no longer put up any convincing front as a boy, so the assumptions of others fit perfectly with where she was in her life. But another U-turn lay just around the next bend.

> I was in the second year of a three-year program when I met one of the boys from my hometown in the hall. He asked if I was going to my former boyfriend's wedding. My face must have fallen a mile, and the tears started to well up. . . . I was totally gutted. Any illusions I had about the future [with this boyfriend] had been dashed, and I simply gave up. I dropped out of school, packed my bags, and went home to my parents' house, back to the prison.

And there Dianne began a downward spiral. She took a job in a nearby town and tried to bury her frustration, her fears, and herself in her work. But none of it helped. And then, one by one, Dianne lost most of her Toronto friends to suicide, or drugs, or prostitution, or all three.

Again she sought medical help. The gynecologists she saw this time told her they could find nothing to explain her condition, though they did confirm that she had unusually low levels of both testosterone and estrogen. She did, they told her, seem remarkably well adjusted for a person in her situation, and they would support her in any decision she made. But there was no one she knew of who could truly offer her any real help with her own particular set of issues.

The spiral continued. At twenty-three, Dianne told her gynecologist she didn't expect she'd live to see twenty-four. One night that November Dianne attempted suicide and failed. "A voice from nowhere said, 'Not yet. Hang on just a little longer.' I wept, and I wept," she said. "Someone or something was telling me there was something worth waiting for."

A few days before Christmas that year, one of Dianne's old Toronto friends called and told her about a doctor in Trinidad, Colorado, who was helping people like Dianne. The doctor's costs were far less than the European clinics Dianne had heard about. The next day, Dianne called the clinic and related her story. The doctor asked to see all of Dianne's records. She sent them the same day.

For the first time, Dianne dared to imagine that she might have a normal life. The wait seemed interminable. She even tried talking to her mother about what was going on, but it was useless. Dianne went so far as to tell her mother about the near suicide. Her mother said it would have been better if Dianne had killed herself. And when Dianne finally said that, regardless of her mother's feelings, she had to go to Colorado, her mother said Dianne could never come home again and could never again speak with family or friends. Her message was clear, but Dianne had made up her mind. If the doctor would have her, she was going.

The doctor finally called and said he would take the case, pending an in-person examination. When Dianne told him she had only half the required money, he paused for a second and said, "Come on down anyway." "That was it!" Dianne said. "I sold everything I owned and left twenty-four years behind with nothing but a suitcase in my hand. I had no idea what lay ahead, but it couldn't be any worse than what was behind!"

On Thursday of that week, Dianne met with the doctor in Trinidad. She explained how she felt, he examined her, and they finalized their decisions. If it was a girl that Dianne wanted to be, then so be it.

The day after Easter Sunday, the anniversary of Christ's resurrection, the doctors operated on Dianne and gave her what she had always wanted, a woman's genitalia. Dianne was reborn.

A week later, Dianne was in Ontario, Canada, and within a few days she had a new job and a new apartment. For her, another life had begun. Here, where no one knew her, Diane could physically and mentally be the girl she had always wanted to be.

I was suddenly and finally free, and I quickly discovered that I was much different than I had ever suspected. I was outgoing, funny, vivacious, sympathetic, and a horrible flirt! As much as I had a taste of freedom in those few years in Toronto, it was nothing like actually *being* free! I suddenly had *lots* of friends, more than my share of suitors, and was truly happy for the first time. I no longer had anything to hide and no limitations. I was a bad girl! I was full of life, sensuous, and wild! It was finally a time when I could experience myself in my totality.

The doctors in Trinidad recommended she follow up with staff at a clinic in Ontario. And for a while, Dianne did regularly visit the clinic. It turned out she was one of very few patients they had ever followed much beyond surgery. The staff was amazed at Dianne, and so was Dianne. Suddenly she was the woman she had always wanted to be—a healthy, happy, normal woman, and that surprised everyone at least a little. For Dianne, all of her former doubts about who she was just disappeared.

"I was simply a sexually and emotionally repressed young woman who had finally cast off the repression and had taken flight. By my thirties my career was on the upswing, and I was living with a man who loved me very much."

At age forty, Dianne finally met her birth mother. At that point much of what Dianne had suspected became real.

I learned some details about the first few weeks of my life and began to get hints that there was something unusual about me at birth. It is likely that I was born with some degree of a DSD (disorder of sex development) that left me neither fully "normal female" nor "normal male," which explains the odd circumstances of my "mixed-up puberty." Whatever the state of my tiny body, it was "corrected" (as much as possible at that time), and I was put up for adoption as a boy.

There is little doubt that many members of the family and some family friends knew of my circumstances, as such knowledge would explain their acceptance of my "difference." It is also likely that this fact accounted for my mother's absolute refusal to accept my expressions of femininity. The "expert opinion" at the time was that nurture shaped the human being, and "raise it as a boy and everything will be fine."

Once the coals of Dianne's suspicions had been brought to full flame, she began to read about intersexuality and the people it affected.

In one of those books Dianne found a picture of a twenty-something person who looked exactly like what Dianne thought she would have if the doctors had not "altered" her at birth.

> I was lightning-struck! I could see *exactly* what had been done—cut here and here, stitch there and presto! . . . Pretty much everything in my early life made sense at that point. I understood why everyone did what they did, and I understood their motivation. I was incredibly angry that nobody told *me*! But at least I understood, and I knew they thought they were doing the best thing for me. Thankfully, I finally defied everyone and did what was right for me before it was too late.

For some time now Dianne has been active in intersex and transgender support groups trying to help others who are struggling with the same issues that she wrestles with. "After all, I have been there, done that," she says. "I may carry the 'scars' of years of mistreatment, misdiagnosis, and misunderstanding, but it also taught me a compassion and acceptance beyond the ordinary that I can extend to others."

Normal Chromosomes Wearing Abnormal Genes

Like a typewriter with a bad key, sometimes a chromosome taps out a story peppered with mistakes. Every time the writer reaches for an "o," the typewriter slaps down an "e." "Looking forward" becomes "leeking ferward." The sense of it blurs and the story takes an abrupt swing.

When a chromosome does that, people change in unpredictable ways.

Besides the genes found on X and Y chromosomes, there are several other genes involved in the sexual development of the fetus. We already know of nearly two dozen such genes, and undoubtedly there are more. A detailed consideration of the enzymes involved in sex development—the nature of each of these enzymes and the corresponding genes, as well as the consequences of mutations—is more than this book can handle. But with so many genes involved, the alternative outcomes of fetal sexual development are like the birds of the air—varied and wondrous. But, also like birds, some are so especially wondrous they deserve a little more attention.

Congenital Adrenal Hyperplasia

Within the urogenital ridge—what will become the external and internal genitalia of the fetus—the gonads and related sexual tissues develop close to the kidneys and the adrenal glands. The adrenal glands, like the gonads, produce hormones critical to determining the sex of the developing child. These hormones include cortisol and the adrenal androgens, especially androstenedione, a precursor to testosterone. When the adrenal glands overdevelop—a condition called congenital adrenal hyperplasia—they can produce enormous quantities of cortisol and adrenal androgens. This can cause otherwise normal 46,XX fetuses to develop clitorises that to varying degrees more closely resemble penises and scrotums—female pseudohermaphrodites.

Mutations in at least five different genes can cause congenital adrenal hyperplasia. And changes in each of these genes cause a distinct set of chemical and physical changes in the child. Baby girls begin to look like baby boys. One of these genes is responsible for an enzyme called 21-hydroxylase. Mutations in this gene result in masculinization of the external genitalia during fetal development.

Interestingly, mutations in the gene for 21-hydroxylase are some of the most common mutations seen in infants. In some populations, such as the Ashkenazi Jews, as many as one in three babies have abnormal 21-hydroxylase genes, as do one in seven people in New York City and one in sixty people in the general population.[13] The symptoms vary considerably from person to person, but genetic heritage clearly plays a role.

Changes in the fetus range from simple clitoral hypertrophy (a large clitoris that may more closely resemble a penis) with normal ovaries, vagina, and uterus, to retention of the urogenital sinus, which is the space where the urethra and vagina usually come together to form a single opening to the exterior of the body. On the other hand, some

46,XY fetuses with congenital adrenal hyperplasia develop like 46,XY fetuses without adrenal problems, and the external genitalia of these infants look like those of a traditional male at birth.

Intersex as a Way of Life Among Other Animals

For those who imagine males and females as *opposite* sexes, hyenas are a conundrum. In fact, the appearance and biology of spotted hyenas have been enough to force some of us to rethink just what we mean by male and female.

Nearly two thousand years ago, Ovid and Pliny the Elder both reported that they noticed something truly extraordinary about spotted hyenas: they all appeared to be males. Spotted hyenas are common in the savannahs and woodlands of sub-Saharan Africa. Apparently Pliny, as had Ovid before him, booked transit from Rome to explore the Dark Continent. What each found when he first looked upon packs of the animals was that every spotted hyena appeared to have a penis and a scrotum. That led Pliny the Elder to propose that hyenas could change their sex—male one year, female the next. Others proposed that hyenas were true hermaphrodites, capable of taking either role in copulation. Even as recently as the twentieth century, Ernest Hemingway and others claimed spotted hyenas were bisexual.

Though Pliny and the others may have erred about the sexual nature and capacities of spotted hyenas, they were right about one thing: spotted hyenas are most unusual mammals. They are not hermaphrodites or shape shifters, nor do all hyenas have penises and scrotums. Spotted hyenas do push the sexual envelope, though, and they do all have the appearance of males. Between the rear legs of every spotted hyena there is a penis-like shaft about seven inches long, and a fleshy sack that looks very much like a scrotum. And in about half the hyenas in every pack, that's just what they are—penises and scrotums. But in the other half, the apparent phalluses are fully erectile, large clitorises (or pseudopenises), and what from a distance appeared to be scrotums are in fact labial-scrotal fusions holding two pads of fatty tissue that look like testes. In effect, half of all spotted hyenas are intersex—40,XX females (forty being the normal number of chromosomes among hyenas) whose genitals look like those of males. So, much like women with congenital adrenal hyperplasia, female spotted hyenas sport a normal female complement of chromosomes but look physically like males.[14]

How does that happen to spotted hyenas?

Unlike most mammals, among spotted hyenas females are dominant. All power passes matrilineally, with females assuming their mother's rank after the mother's death.[15] After a kill, males don't eat until the females have finished. Laurence Frank at the University of California at Berkeley described a situation in which one juvenile female held five fully grown adult males at bay while she had her fill from a buffalo carcass.[16] In fact, except during copulation, female spotted hyenas completely dominate males.

To ensure their dominance, as female hyenas are developing they get an overdose of male hormones—in particular the male androgen androstenedione. In other mammals, including humans, during female development much of this androstenedione gets converted into estrogen. But in spotted hyenas most of the androstenedione becomes testosterone. And the higher the female hyena's status, the more testosterone she passes to her offspring, both male and female, but especially to the females. All of that testosterone, along with some other, less well understood contributors including genetic factors, masculinize developing female fetuses in several ways. Females do develop vaginas and uteruses, but their clitorises enlarge to equal the reach of male penises, and the labia become a little more like scrotums and never fully separate as they do in many other female mammals.[17] Body size increases, as does aggression.

An interesting experiment in mice, pigs, rabbits, and other litter-bearing species supports the idea that as female spotted hyena fetuses bathe in testosterone they become more aggressive. During the development of a litter it is possible for a female fetus to develop in the uterus situated between two male fetuses (2M), between a female and a male fetus, (1M), or between two female fetuses (2F). It turns out that if you are female, who you sit next to *in utero* makes a difference. A 2M female produces fewer living offspring, has higher blood levels of testosterone and larger scent glands, and is much more aggressive than 2F females—all apparently because of testosterone spillover from her littermates. There may be some evolutionary benefit to this. Among these same litter-bearing species, during times of high population density, the most aggressive females are also the most likely to reproduce in spite of their reproductive problems. That makes it appear that when food gets scarce the most aggressive females are the most successful. That would be an enormous advantage to any female living in a place where food is never abundant. And all of it arises simply from sharing a placental bed with a brother or two.

Among female spotted hyenas there is a hierarchy. Alpha females are the most aggressive of all the hyenas, and while they are pregnant they give their fetuses the largest doses of testosterone. That testosterone ensures that her offspring will be equally aggressive and dominating. But there is a price; female spotted hyenas' ovaries are partially stunted, increasing the difficulty of conception. In addition, the females' genitalia look a lot more like those of males, including the enlarged clitorises, the labioscrotal fusions, and the failure of the ureter to separate from the vaginal opening, creating a long penislike vagina beneath the hypertrophied clitoris. This also increases the difficulty of successful mating. Among the four species in the hyena family, only spotted hyenas are built like this.

So how do spotted hyenas reproduce? With a lot of practice. Young males often struggle to successfully penetrate a female spotted hyena. Only through years of repetition do they get good at it. Even then the female has to provide a lot of cooperation. Appearances to the contrary, female hyenas' pseudopenises differ considerably from penises both functionally and anatomically. During copulation the female's "penis," or enlarged clitoris, serves as the opening to her vagina. When the male is properly positioned the female can invert her clitoris, pulling it up inside of herself like turning the sleeve of a long-sleeved T-shirt outside in. That gives the male access to her vagina and creates the possibility of fertilization and pregnancy.

Though successful fertilization is the key to hyenas' future, for the female it can be an unpleasant prospect. At birth the baby hyena enters the world through the same pseudopenis that the hyena uses for urination and copulation. That requires a lot of changes.

"During birth it [the pseudopenis] stretches out so much that it looks like a water balloon."[18] And sometimes that balloon bursts. In fact, during first pregnancies the clitoris must break open for successful delivery. After that some females never become pregnant and give birth again. Others die. Even when everything goes right, or nearly right, with the mother, in captivity about 60 percent of hyenas are stillborn.

That is a trade-off for the powerful, dominant females that have made the spotted hyenas as successful as they are today. The unusual development of female spotted hyenas provides the edge they need to compete successfully, even against males, for too little food.

Spotted hyenas have been around for about seven million years, all that time hunting prey, mating, reproducing, raising their young, and surviving in a harsh environment. And all that time the females have had an anatomical "problem" that, if they were humans, we would insist on fixing. For spotted hyenas, intersex is simply a way of life.

Androgen Insensitivity Syndrome

Other genetic changes that affect androgens can lead to male babies that look like, or very nearly like, baby girls.

The process that leads to the formation of penises and scrotums during fetal development requires not only the production of androgens by both the testes and the adrenal glands but also the normal expression of receptors for these hormones. That is, the hormones—like testosterone produced by the testes and the adrenal glands—normally bind to specific molecules on the surface of cells in the developing genitalia, like a key slipping into a lock.

When the lock and the key come together just right, a baby boy begins. When the androgens (the keys) are present but the receptors (the locks) have changed, nothing happens. This condition is called complete or partial androgen insensitivity, depending on the degree of receptor change or loss. In androgen insensitivity the developing tis-

Hormones bind to specific molecules on the surface of cells in the developing genitalia.

sues—even though bathed in androgens—cannot use these hormones. Then 46,XY babies are born that cover the range of possibilities from nearly normal baby boys to almost perfect baby girls, even in the presence of a normal Y chromosome with SRY and normal levels of male androgens. The gene responsible for the normal production of androgen receptors is called AR, and it is on the X chromosome.[19]

Because AR resides on the q arm of the X chromosome, losing pieces of the q arm can have dramatic consequences during sexual development. Nearly one hundred different known mutations in AR cause partial or complete androgen insensitivity.[20]

Even though their sex chromosomes are normal, these babies have been called pseudohermaphrodites because of the disconnect between their sex chromosomes and how they appear to us.

NICKY PHILLIPS
AGE: 64
LOCATION: BRITISH COLUMBIA, CANADA

"Recently I had to fill out a form for Census Canada. I found one of the questions to be problematic. It wasn't one of those hard ones where you were expected to calculate your income or the time you spent taking care of seniors, it was the one which asked, 'Are you male or female?'"

Nicky is a smart, fast-thinking, and thoughtful woman living in western Canada. She is sixty-four years old and retired. She is also

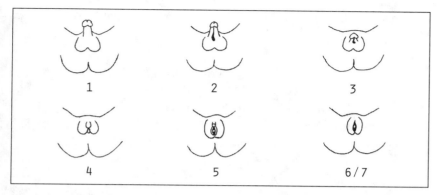

Schematic representation of a grading scheme for clinical classification of AIS.
Grade 1: normal masculinization in utero
Grade 2: male phenotype with mild defect in masculinization
Grade 3: male phenotype with severe defect in masculinization
Grade 4: severe genital ambiguity
Grade 5: female phenotype with posterior labial fusion and clitoromegaly
Grade 6/7: female phenotype[21]

pretty good with this sort of question. She has to be. Nicky is a woman with complete androgen insensitivity syndrome.

"I checked female, which is the sex on my birth certificate and which is the closer to correct of the two, but I was tempted to fill in both squares. It would have been interesting to add the question, 'Are you talking chromosomes or appearance?'" Nicky offers.

When Nicky was born in 1943, the doctors told her parents that she was a girl, but there was a slight problem. Nicky, they said, had a double hernia that they would have to repair in the near future. In 1943, essentially nothing was known about fetal sexual development and the world of possibilities that it offered. No one, I believe, suspected that Nicky's apparent hernias were actually partially descended testicles.

So eighteen months after she was born, when the doctor found something he didn't expect in Nicky's "hernias," something he imagined to be ovarian tissue, he simply placed the tissue where he thought it belonged, back in her abdominal cavity, repaired what looked a little like a pair of hernias, and stitched her up.

> Adolescence was a difficult period for me. . . . At sixteen, I was told that I had been born without a uterus and that I would not be able to have children. When other girls at school would discuss their periods, I became silent. One of my worst experiences in high school occurred when my PE teacher said that we would all have to shower together in the "gang showers." One of the characteristics of AIS [androgen insensitivity syndrome] is an absence of secondary hair. I was mortified that my classmates would see that I had no pubic hair. As it turned out, another girl and I were the last ones to complete a PE exercise and we were able to get into the private showers. The other girls tried to peek in, but I was able to keep facing the shower heads until they left. This episode didn't enhance my popularity at school.

Most of us can only imagine what Nicky went through—the confusion, the shame, the fear about what seemed so out of the ordinary, so abnormal compared to all of her friends. And it would be nine more years before Nicky would pull together all the pieces of her story and begin to understand what was taking place inside of her body.

Before enrolling in college, Nicky had to undergo a physical exam with her family doctor. That doctor told Nicky that her vagina was shorter than normal. He didn't mention any problems that might create during attempts at sexual intercourse. But Nicky and her boyfriend were already experimenting with sex, and she had discovered on her own that sex for her seemed to be a somewhat different experience than it was for others.

It took a visit to a second physician to find out that vaginal reconstruction, or vaginoplasty, was a possibility. This second general practitioner referred Nicky to a gynecologist. That referral would change the rest of Nicky's life. Then twenty-five years old, Nicky was about to uncover the final clues about herself. The gynecologist told Nicky that she had a condition called "testicular feminization," which is another name for androgen insensitivity syndrome and was the only name for this condition until the roles of androgens and their receptors became known. Androgen insensitivity syndrome may be partial or complete. Partial androgen insensitivity syndrome comes in lots of different forms, each of which comes with its own particular set of changes. Because Nicky had fully female genitalia, she had what is now known as complete androgen insensitivity syndrome. As many as 1 in 20,000 babies born every year have complete androgen insensitivity syndrome. In the United States, that's about two hundred babies and about twelve hundred worldwide every year.

Nicky's gynecologist suggested a vaginoplasty to enlarge her vaginal canal, and he made one other suggestion. Along with the plastic surgery, he thought it might be a good idea to perform an exploratory laparotomy to see if Nicky still had any gonads in her abdominal cavity and, if so, whether those gonads were testes or ovaries.

During the laparotomy, Nicky's doctor found what he had imagined he would find—a pair of testes. He removed the testes and remodeled Nicky's vagina.

"I did not realize, then, how fortunate I was to have been referred to that particular gynecologist," says Nicky. "It seems that the conventional wisdom of the time was not to tell patients that they had male chromosomes for fear that the information would cause them to lose their sanity. Many people with AIS with whom I have come in contact searched for years until they got a diagnosis."

Another woman with complete androgen insensitivity told me that at age fifteen, her parents and her doctor told her she had to have her "cancerous" ovaries removed. In fact, she was having an orchidectomy—the doctor was removing her testes. For years afterward this woman had serious (and groundless) fears about cancer, but no one would tell her the truth. Years later, and then only by accident, this woman finally discovered the truth about herself.

Nicky's experience was better than some. She notes, "On the whole, I feel I was treated pretty well by the medical profession. Dilation was not a part of a vaginoplasty in 1968. I went through a period of sexual inactivity and lost some vaginal length. As it happens, I am now in a lesbian relationship, so I don't see this as a major concern. One interesting note," she continues, "the gynecologist I saw was a very shy man; it took my GP to ask me if I was experiencing orgasm."

Nicky finally realized that this was more than she could face alone. In the end, she turned to therapy. "At twenty-five, I thought I had dealt with it, and I had a good intellectual understanding of my condition." It wasn't until four years later that the emotional aspect hit her. "I had to deal with feelings of inadequacy about not being a 'real' female. [And I had to] address my feelings around not being able to have children. The therapy helped a lot, and I would like for such help to be made available to all young girls with androgen insensitivity syndrome."

When I asked Nicky what she would like for others to know about her, she said:

> I am financially secure and in a stable relationship. My spouse and I were married in 2003, when same-sex marriages became legal in British Columbia. My spouse had her gender reassigned [from male to female] in 1992, just prior to my meeting her.
>
> We live in a townhouse, get along reasonably well with our neighbors, and are very similar to other retired couples. I would like society in general to be able to grow up a bit around issues of sexuality and gender. I happen to have AIS, which is a condition passed down through the X chromosome. If it were color blindness, which is passed the same way, no one would think it was a big deal. Because it is tied in with issues of sexuality and gender, it becomes something to be laughed about or discussed with discomfort.
>
> I remember having to be very quiet and careful in high school when the conversation turned to menstrual periods. I would really like it to be safe for young women with AIS to disclose their conditions.
>
> I do believe there is a naivete about the way we ignore the fact that some people don't fit neatly into the either-or of gender. I believe that gender is rather a continuum than an either-or proposition.

Children Who Change Their Sexes: 5-Alpha Reductase Deficiencies

Some researchers refer to male and female pseudohermaphrodites, people with AIS for example, as sex reversals, meaning that the sexes of the external and internal genitalia do not correspond. This, of course, is not sex reversal, it is simply another situation that does not fit very tidily into a two-sex world. There is, however, one enzyme deficiency that leads to something that *is* very nearly a sex reversal, at least in the minds of those who experience it.

Normally, the final strokes in the creation of a baby boy require two hormones, testosterone and a derivative of testosterone called dihydrotestosterone. The developing testes produce testosterone, but an enzyme derived from another source is necessary for the conversion of testosterone to dihydrotestosterone. That enzyme is 5-alpha reductase type 2, normally produced from the SRD5A2 gene at relatively high levels during the sexual development of boys.

But nearly thirty known mutations in the SRD5A2 gene result in inadequate production of functional 5-alpha reductase type 2. When that happens, 46,XY fetuses develop normal testes, but the testes don't descend, and their genitals look like those of baby girls. As a result, at birth they are almost always identified as baby girls. And they nearly are, except they have testes, and what seems to be a vagina is only a small closed pouch that leads nowhere. Only a few of these children have sufficiently ambiguous genitalia that physicians recognize their condition at birth. So most often the first words baby boys with 5-apha reductase type 2 deficiency hear are, "It's a girl." And usually that's the way it remains until puberty—the child lives its life as a girl. All the while, though, a storm is brewing, and at puberty that storm erupts in all its fury.

As sexual maturity approaches, another form of 5-alpha reductase called type 1 appears along with a big boost in testosterone. Perhaps it is the addition of the other form of 5-alpha reductase and/or all the testosterone that causes the changes. Regardless, the results are remarkable. What everyone thought was a clitoris begins to grow to as much as three inches long, the voice deepens, muscles enlarge, and the owner of that brand-new almost-penis begins to have sexual feelings for others, feelings that can cause erections. And apart from the physical changes, there is the psychosocial upheaval that these people must experience. A girl until age twelve or thirteen abruptly becomes mostly a man.

Fishy Sex: Changing Times

To most of us, a girl becoming a boy is astounding. Not so with fish.

Among vertebrates, fish dominate. With some thirty thousand species ranging in size from about one-third of an inch long (the Philippine Island goby) to about fifty feet long and weighing several tons (the whale shark), no other vertebrate species compares to fish for variety and sheer numbers. Among those of us with backbones, there are more fish than any other creature. And when it comes to sex, fish have evolved some of the most varied and interesting approaches among all the animals.

More than (perhaps a lot more than) one hundred species and twenty families of fish are hermaphroditic, and here we begin to stretch the limits of what we mean by hermaphroditism. Hermaphroditic fish come in two common forms—simultaneous hermaphrodites and sequential hermaphrodites. Simultaneous hermaphrodites have the reproductive organs of both sexes at the same time. Sequential hermaphrodites have ovaries for parts of their lives and testes during other parts of their lives. Hamlet fish are simultaneous hermaphrodites and have both female and male sexual organs as adults. The same is true for some types of salmon. Hamlet fish are small, gold and yellow fish found mostly in the Caribbean and the Gulf of Mexico. And though they have both male and female organs, these fish do not mate with themselves. Perhaps this is simply a consequence of their shape and size, or maybe it's an evolutionary adaptation that helps maintain or generate diversity among hamlet fish. Regardless, when hamlet fish do mate with one another they take full advantage of their hermaphroditism. During mating, hamlet fish take turns being the male and the female partner. Hamlet fish trysts involve multiple matings that last for up to three nights. So each fish has several opportunities to try out the role of each sex. For such small fish, their lust, not to mention their creativity, is great.

Some hermaphroditic sea bass, on the other hand, do in effect mate with themselves. These bass may spawn as many as twenty times in a single day. And as they spawn they alternate between being egg-laying females and sperm-spouting males. It takes a sea bass only about thirty seconds to go from male to female, from laying eggs to fertilizing those same eggs with a dose of bass sperm. Because sea bass have both ovaries and testes, these animals are by definition simultaneous hermaphrodites. But they don't function as males and females simultaneously; rather, they switch from one to the other with surprising speed and seemingly at will. And as these fish age, things get even more unusual.

As sea bass pass into their golden years, their ovaries enlarge, and they begin to produce mostly eggs and little to no sperm. That seems to increase their reproductive rate, since most eggs do get fertilized (by other fish) while most sperm do not find eggs. Thus, in effect, these sea bass end their lives as females.

Other species of fish do just the opposite and end their lives as males. Because they are not true sequential hermaphrodites, fish like these sea bass are sometimes called successive hermaphrodites, meaning there is one point in their lives when they do not have organs of both sexes and one point at which they do. Wrasses (brightly colored fish often seen in saltwater aquariums), parrotfish (nearly iridescent fish found widely distributed among oceanic coral reefs), and some gobies (small, cigar-shaped fish that make up the largest group of marine fishes) are all also successive hermaphrodites.

These fish find themselves beyond the sexes for one phase of their lives—not exactly female or male—but end their lives as more or less males or females. They are fish with a flair for change, fish with no concept of sex as we think we know it. Whole schools of fish fool with our heads and confound any simplistic idea we might like to have about sex. But the story gets even better.

For decades, maybe centuries, people have known that some fish change sexes during their lifetimes. But it wasn't until 1972 that marine biologists began to figure out what motivated these fish to up and abandon their lives as males or females and sprout the genitalia of the opposite sex.[22] Not surprisingly, it turns out that the whole motivation thing is complicated.[23] Every fish seems to have its own set of rules and reasons for swapping sexes.[24]

Beyond successive hermaphroditism, some wrasse (*Labroides dimidiatus*) are sequential hermaphrodites—at times female, at other times male. Most of these small coral-reef fish begin life as female. And as they grow, a complex social structure develops so that by the time these fish achieve adulthood they live in harems of female fish controlled by a single dominant male wrasse. This alpha male, through his physical domination and perhaps his chemical presence, forces the females to remain female. Among the females of the group a pecking order is quickly established, and the alpha female runs the show. But her most important job is yet to come. When the one male wrasse dies, over the course of a few days the alpha female becomes the alpha male and takes over. From veiled damsel to bearded sheik, from a maker of eggs and a receptive mate to a sprayer of sperm and the master of all in a day or two—this is sex as a choice, sex as a consequence, sex as social order.

Among clownfish—small orange- and white-striped marine fish edged in jet black—a similar transformation takes place, but with a slightly different twist. Again, these fish assemble themselves into groups made up almost entirely of females. But among clownfish, only the largest female in the group can mate with the single large alpha male. If the large female clownfish dies, the big male becomes female. After that, the largest of the young females leaves her egg-laying days behind, develops testes, and becomes the alpha female. So among clownfish, the few and the proud begin life as females, swap gonads for the grander life of the leader of the pack, and then—for the greater good—reclaim their ovaries and lay eggs as sweetly as any clownfish who'd ever graced this Earth. A tripartite tryst with a sexual subtext unlike any human ever imagined. Once more the idea of two sexes, especially two opposite sexes, seems strained by the reality of living animals.

Reef gobies—lacy-finned, hand-sized fish—take an equally eclectic view toward the utility of sex. Again, when the dominant male dies, one of the females in the group will become the dominant male. But if a larger male should happen by and choose to take an interest in the females, the once-female-now-dominant-male fish will re-create her ovaries and live out the rest of her life as a reproductive female—unless, of course, the new male also dies. If that happens, then a whole new round of sex changes begins. And none of these changes seem to have any lasting effect on the fish. Certainly there are obvious differences between male and female gobies—color, gonads, hormones, etc.—but even repeated changes of sex do not seem to have any lasting effect. A goby that becomes a female for the first or the third time is every bit as much female. The same is true for males. The only apparent downside is that changing sexes requires a great deal of energy—lots of new proteins to be made, lots of new hormones, dramatic increases in enzyme activity, pigments to be laid down, etc. So, while the animals are changing sexes, it seems likely that they are easier targets for predators. But they do survive to offer us a whole new window into the world of sexuality—sex change as conflict resolution.

And then there are species like the saddleback wrasse, midshipman fish, and some species of salmon that have two very different types of males. Saddleback wrasse are brilliant blue fish about a foot long with an orange-brown "saddle" just behind their heads. They live, among other places, in the waters off Hawaii. In schools of saddleback wrasse there are two very distinct types of males—one large, one small; one very reproductively active, one less so; one with a lot of brain cells that make arginine vasotocin (AVT) neuropeptide, and one with very few cells that make AVT. Arginine vasotocin neuropeptide stimulates mating behavior in a variety of vertebrate animals. The genetic differences (if any) between these two types of males remain unknown. But the differences between their roles in saddleback wrasse and other fish societies are obvious. One of the males spends a much larger portion of his life wooing lady wrasse and sowing his seed. The other male lives a more sedate life and knows fewer females than his more active counterpart. On the surface that would seem to be evolutionary suicide, but the less reproductive males still swim among the wrasse just as it appears they have for millennia. Both males must serve the wrasse society in important ways, because if the smaller, seemingly less reproductive wrasse were nothing more than evolutionary aberrations they would have shuffled off this mortal coil eons ago.

Midshipman fish live off the west coast of the United States, and the differences between male midshipman fish further stretch our ideas about two sexes. Biologists have named the two different sorts of male midshipman fish type I and type II. Type I midshipman males take the longest to mature. During those extra days and nights, type I males grow larger and louder. Midshipman fish use a limited series of vocalizations to attract mates, so the louder a midshipman fish, the better chance he has of mating. Type II males take the energy that type Is put into size and song and use it to inflate their testes. As adults, type II males' testes make up almost 10 percent of their body weight. That's comparable to a 175-pound man having 17.5 pounds worth of testicles—about the weight of good-sized bowling ball. These fish have just one thing on their minds. Type I males, on the other hand, devote only about 1 percent of their total weight to testes—not trivial (think of a 175-pound man with nearly two pounds' worth of testicles), but nowhere near what the type II males do.[25]

The two types of males also secrete two different types of testosterone, and only type I males make and guard nests. Inside the warmth of those nests, these guys hum Johnny Mathis tunes with their husky voices and wait for females to come to them. Type II males don't sing and don't build or guard nests. Instead, these Lotharios prefer to slip into a type I male's nest and mate with the resident female while hubby is away. Just what evolutionary advantage this offers to midshipman fish isn't clear. But among the saddleback wrasse and the midshipman fish, three very different sexes persist.

No matter how hard we may try to squeeze these fish tales into our human stories, sex remains—as pioneering geneticist J. B. S. Haldane said of all the universe—"not only queerer than we suppose, but queerer than we can suppose."

6

WHAT WE DO ABOUT THE
AMBIGUOUS CHILD

Dozens of species of fish do it. Even female spotted hyenas with penises do it. All of these creatures mix in more than two sexes, move from female to male and in between as easily and quickly as we change clothes, and that doesn't seem to bother any of us. We don't feel that we ought to *do* something about these animals. But when it comes to human beings, everything changes. We have social norms, expectations, and high school showers to deal with. And of course we have the vise of language with its steel pincers. When it comes to human intersex, many people find themselves in a curious and frightening void. The only way out of that void is through the child.

Limiting the Ambiguity: Assigning a Gender

In the summer of 2006, the Lawson Wilkins Pediatric Endocrine Society and the European Society for Paediatric Endocrinology gathered together "fifty international experts in the field" to answer some of the questions that surround the birth of an intersex child. Their report appeared in 2007 as the "Consensus Statement on Management of Intersex Disorders." This group was not the first, and I'm certain it won't be the last, to tackle the thorny issues of sex and gender assignment, the welfare of the children and their parents, what sort of expertise should be available, and how physicians can best serve all of those involved. It isn't entirely clear why this task should fall to physicians, but I think their conclusions offer a useful insight into some of our current thinking about intersex.

First, this group recommended that we change the words we use to describe intersex people. "Terminology such as 'pseudohermaphroditism' is controversial, potentially pejorative to patients, and inherently confusing." They proposed in its place the term "disorders of sex development" or DSDs "to indicate congenital conditions with atypical development of chromosomal, gonadal, or anatomic sex." That seems a reasonable conclusion. After all, very few of us, even without years of medical training, feel that the term *hermaphrodite* adds to our social currency. And this conclusion acknowledges, up front, the crucial role played by language in this whole area. The language we use to describe ourselves and the words that others use to name us change the way we see everything. Words are not innocent bystanders here.

Then it was time to address the hard clinical issues, gender and sex assignment, surgeries, expertise, parents, and children.

"Standards of care for best clinical management of DSD include a gender assignment for all; avoiding gender assignment before expert evaluation in newborns; open communication; multidisciplinary-team evaluation and management; family/patient participation in decision-making, respect and attention to patient/family concerns; and strict confidentiality."

"A gender assignment for all." What that really means is everyone, and we mean everyone, has to be either a boy or a girl. If we have done away with hermaphrodites and pseudohermaphrodites, what else is there? And those involved must choose between these two options just as soon as the team and the parents can agree on what's best for this child, the consensus being that no one in our society can wander around productively and sanely without an attached indication of sex.

And who should make those decisions and should assist?

"The core team should consist of pediatric endocrinologists, surgeons, urologists, or gynecologists, psychologists/psychiatrists, geneticists, neonatologists, social workers, nurses, and medical ethicists."

"DSD should be suspected 1) in cases with overt genital ambiguity (such as cloacal exstrophy [a DSD where no genitalia develop and the lower abdomen fails to close], 2) if the genitalia are apparently female but include an enlarged clitoris [This particular study doesn't say exactly when a clitoris should be considered enlarged. However, it does provide very specific information about penis sizes in Japan, India, Australia, the United States, and Europe—from 3.5 cm in the United States to 3.6 cm in India at birth], 3) apparent male genitalia with bilateral undescended testes, micropenis, isolated hypospadias (when the urethral opening falls somewhere below and behind the tip of the penis), or mild hypospadias with undescended testis, 4) a family history of DSD, or 5) when karyotype and physical appearance disagree."[1] A much earlier study did deal with issues of clitoral size and the limits of normalcy. In this report the author devised what he called the "clitoral index," a number obtained by multiplying width of the glans by the length of the phallus. According to this study, a clitoral index of less than 3.5 is normal and a clitoral index of greater than 10 is cause for concern. So we do have a set of numbers to refer to for help in estimating the normalcy of either a penis or a clitoris.[2]

Therefore, careful physical examination, karyotyping, and blood work are all recommended whenever DSD is suspected.

There have been a surprisingly large number of papers and study groups offering advice on how to identify and deal with newborn children with sexual ambiguity. In the recent past, at least, most of them

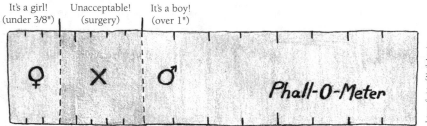

This Phall-O-Meter created by the Intersex Society of North America shows the current medical standards for children born with ambiguous genitalia.

have reached these same basic conclusions—be thorough and use every piece of information available to help with the decision. This seems reasonable. After all, if a newborn child shows any sort of physical abnormality, any sane person involved would want, as soon as possible, to know what was unusual about this child and what might be done about it. But beyond the approach and the speed of the evaluation, there is much less agreement among physicians.

Once, this wasn't a problem. Before the early 1990s, physicians often chose to keep parents in the dark about their unusual children. This tactic was an offshoot of Dr. John Money's powerful influence on perceptions of human sexuality. Money believed that the final sex of any child was determined primarily, if not solely, by the child's upbringing—all, or very nearly all, nurture, not nature.[3] The issue for the pediatrician was to determine the optimum sex of rearing, regardless of the genetic sex. Determining the optimum sex could involve any number of things, including, with older children—according to one of Dr. Money's patients—screening pornographic movies and observing the patient's response.[4]

The less they knew about what the physicians were up to, the less likely the parents were to interfere with the doctors' visions for the children. If nobody but the doctor knew, girls would never suspect that they were boys or vice versa.

By the beginning of this century, some physicians (along with most parents and patients) had begun to advocate for a more reasoned approach. For example, in a paper by Drs. Jorge Daaboul and Joel Frader, the two physicians proposed a new terminology and a transfer of more of the physicians' powers into the hands of parents and, when possible, the affected children. As these two doctors stated, too often "the 'tradition' of sex and gender assignment reflects physicians' preferences, custom, or even bias. The physician-centered approach favors reproductive potential over ease of intercourse or sexual pleasure in the overvirilized female; and ease of intercourse trumps reproductive potential in undervirilized males. . . ." They concluded, "The traditional medical and surgical approach to newborns with intersex maintains a morally and legally unacceptable paternalism."[5]

The idea that the physician was the best qualified to make life-shaping decisions for families and children foundered but didn't die.

The next step in the evaluation of the potential intersex patient, according to the doctors involved in composing *The Consensus Statement on Management of Intersex Disorders*, is to assemble the core team (of physicians) and assign a sex. Assigning a sex is complicated, and the final choice depends greatly on the nature of the DSD identified. In general, though, this consensus report suggests that these decisions should be based on studies of the satisfaction of individuals who have had similar experiences.[6] In other words, try to find other intersexed individuals who were raised as a boy or girl, for what reasons, under what circumstances, and then figure out who ended up happy and who did not.

For example, according to one set of studies, more than 90 percent of 46,XX people with congenital adrenal hyperplasia (which causes the development of nearly male external genitalia) and all patients with complete androgen insensitivity (which causes 46,XY fetuses to develop nearly female external genitalia) who were assigned as females in infancy seem happy as adult females. Therefore, all kids who are 46,XX with congenital adrenal hyperplasia or who are 46,XY with complete androgen insensitivity syndrome should be assigned as females, and so on. Furthermore, according to this study, sex assignment should happen as quickly as possible after a full evaluation by the medical team.

Notice that to this point nothing has been said about surgery. And some, perhaps many, believe that nothing needs to be said at this point. According to a 2004 report, "All [involved] should understand that a decision to raise an infant as a boy or girl in no way depends on surgery, *per se*."[7] And the 2007 consensus similarly concludes that, even after assignment of a sex, "appearance-altering surgery is not urgent."

This seems, at least in part, to be a concession to the efforts of a large number of people, including intersex people like Cheryl Chase (born Brian Sullivan) who have campaigned for years to educate people about the realities of intersex persons and surgery.[8]

Still, the next consideration according to the 2007 consensus guidelines is surgery: "Rationale for early reconstruction includes beneficial effects of estrogen on infant tissues, avoiding complications from anatomic anomalies, satisfactory outcomes, minimizing family concern and distress, and mitigating the risks of stigmatization and gender-identity confusion of atypical genital appearance. Adverse outcomes have led to recommendations to delay unnecessary genital surgery to an age of patient informed consent, although the relative risks and benefits are unknown. . . . The goals of genital surgery are to maximize anatomy to enhance sexual function and romantic partnering."[9]

Another source refers specifically to feminizing surgery: "The immediate goal is to provide the external genitalia with an esthetic and feminine appearance. The long-term goals are to produce a functional vagina of sufficient size for sexual intercourse, to retain sexually sensitive tissue to allow orgasm, and if internal genitalia permit, to preserve fertility potential."[10]

Choosing to raise an intersex child as either a boy or a girl is important, but for some it is not enough. Legitimate fears of stigmatization and social isolation, coupled with the greater likelihood of success of some surgeries during infancy, drive many families and doctors to seek early surgical solutions to sexual ambiguity. Add to that the mythology that surrounds so much about sex in our society, and it becomes easy to understand why the solutions to sexual ambiguity still frequently include surgery.

At that point the question becomes not if, but how? The problem now is how to choose among literally dozens of surgical approaches.

Feminizing Surgeries

The single most common cause of genital ambiguity is congenital adrenal hyperplasia (see page xx). When enlarged adrenal glands begin to pulse the developing fetus with testosterone, 46,XX females start to look, to varying degrees, like males. The most common consequences include an enlarged clitoris and a less than fully developed vagina. Most parents and physicians opt for surgery.

The two procedures most commonly performed to eliminate these "abnormalities" through genitoplasty (reconstruction of the genitalia) are clitoroplasty—clitoral reduction/resection, and vaginoplasty—construction or enlargement of a vagina.

Clitoroplasty seems to be the simpler and the more uniform of the two procedures. First the surgeon cuts and peels away the skin surrounding the shaft of the phallus, leaving only a small strip of skin connecting the skin on the head and at the base of the phallus. He or she then removes a pie-shaped section from the head of the phallus and sutures the remainder back together to form a smaller, more clitoris-like structure with hopefully at least some of the sensitivity of the glans intact. Then, to shorten the shaft, the surgeon cuts out a portion of the middle of the shaft of the phallus, discards it, and sutures the remaining pieces back together. The completed structure can then be slipped into what is or will become the child's vulva—if need be, constructed from the remains of the skin peeled away from the penis and a portion of the scrotum.[11]

Vaginoplasty is an entirely different matter. A partial listing of techniques for vaginoplasty includes: the posterior omega-flap method, total urogenital sinus mobilization method, the posterior sagittal approach, the posterior sagittal pararectal approach, vaginal tract without skin graft, the split-thickness skin graft vaginoplasty, skin-flap vaginoplasties, bowel-segment vaginoplasties, amnion vaginoplasties (in which a section of human amnion from the amniotic sac is used to create a vagina), peritoneum vaginoplasties (in which a portion of the lining of the peritoneal cavity is removed and rolled up to create a vagina), bladder mucosa vaginoplasties (in which the bladder is filled with saline, the muscle opened to the level of the mucosa, and the mucosa removed, filled by an inflatable mold and inserted into a prepared space), scalp, buccal mucosa, and fetal-skin vaginoplasties.[12]

Even a tomish textbook might not offer space enough to discuss all of these procedures. I'll describe only a few of the more commonly used approaches for construction of a vagina in a child with either congenital adrenal hyperplasia and partial or complete androgen insensitivity syndrome—the two most common disorders of sex development that end in surgery. In most children with congenital adrenal hyperplasia, there is at least a rudimentary vagina, though it often does not open directly to the outside. In other children there is no vagina at all. Either way, it is possible to create one.

When there is no vaginal rudiment, a vagina can be fashioned from a piece of colon (large intestine) or ileum (the terminal section of the small intestine). A section of bowel offers the advantage that it is already cylindrical and approximates the structure of a vagina. On top of that, these tissues secrete mucus, somewhat like a normal, functioning vagina. The surgeon simply cuts out the section of bowel and inserts it into a hole cut between the clitoris and the anus, sews the section of bowel in place, and creates a vagina. Usually this turns out to be only the beginning, and at least one more surgery must follow to construct a vagina adequate for sexual intercourse.

In baby girls who have severely enlarged clitorises and rudimentary vaginas, clitoroplasty itself often provides the material necessary for creating a vagina. One commonly used procedure begins with "degloving" the enlarged phallus—in effect, pulling the skin from the "penis" like peeling a banana—except for the skin at the tip or glans.

The flaps of phallus skin are then sewn into a tube still attached to the base of the phallus. The surgeon turns the tube inside out, pushes it through a hole opened near the rectum, and stitches the new vaginal tube to the rudimentary vaginal tube naturally present. Clitoroplasty is then performed to reduce the size of the remainder of the phallus and to create a more nearly normal clitoris. Finally, the scrotum or labioscrotal tissue is manipulated to construct nearly normal-looking labia.[13] At the end, like a canyon carved into human bedrock, the new clitoris sits, as it should, at the apex, just above

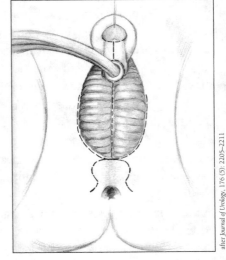

after *Journal of Urology,* 176 (5): 2205–2211

Planned incisions for combined clitoroplasty and vaginoplasty.

the vaginal opening, both surrounded by the red-rock folds of the labia minora and majora. Again, though, the surgeon's work is likely incomplete, and more engineering and reengineering will have to happen before such a vagina can achieve its full potential.

Much of the potential pleasure of adult intercourse lies inside that final bit of phallus still rooted in its native soil. But following most of these surgeries it isn't clear if what remains will still provide pleasure. [14]

Some people, less changed physically by congenital adrenal hyperplasia, can learn to enlarge their vaginas by self-dilation. These children begin life with a more nearly normal vagina that opens to the exterior of their bodies, and they can learn to enlarge their own vaginas by inserting an increasingly larger series of tampon-like tubes that stretch the tissues and slowly force the vagina to expand—a sometimes painful but also sometimes effective approach to reshaping the opening.

Many babies with congenital adrenal hyperplasia have additional issues to contend with. During fetal development, the tissues that will eventually form the bladder, the urethra, and the vagina begin as a single unit called the urogenital sinus. Usually, as the fetus rounds out its development, the vagina, urethra, and bladder separate from one another and seek out their own geographical spots on the pelvic map. When that happens, the urethra separates completely from the vagina and forms a separate tube that connects to the outside, just below the clitoris.

In many children with congenital adrenal hyperplasia, that separation never happens. The result is a single external opening that divides internally into two passages, one of which leads to the bladder and the other to the vagina (or rudimentary vagina).

Traditionally physicians said these children had a "high vagina" if the vagina joined the urethra before the external sphincter (the one we become most aware of using after drinking a quart of beer) and a "low vagina" if it joined the urethra after the external sphincter. The problems with that approach are the same as those inherent in the two-sex idea. The words don't really tell us much about the range of conditions that exist in between and beyond "high" and "low" vaginas.

Others have tried to eliminate the ambiguities of the high-or-low model by certain quantifiable measurements.[15] But all attempts to reduce genitalia to numbers for the convenience of communication have met with only limited success. The beauty, the breadth, the width, the length, and the abnormality of it all remain largely in the eye of the beholder.

But the presence of an incompletely developed urogenital sinus has led to some remarkable surgeries to reconstruct our children. One of these procedures, called total urogenital sinus mobilization (TUM), requires the dexterity of a puppeteer moving the strings on a dozen marionettes at once.[16]

To separate the urethra and the vagina in children with a retained urogenital sinus, one catheter is inserted into the bladder and another into the vagina. After that, the surgeon makes an incision just below the phallus (clitoris) to just above the anus. Then the whole urogenital sinus (including the uterus) is freed by dissecting it away from all of the surrounding tissues.

Using these two catheters like the strings on marionettes, the surgeon then manipulates both bladder and vagina into their scripted spots on the stage and sutures them into place.

It's a technical miracle, and where there was doubt there is now a girl. But this surgery is not without risks. And even as complicated as all of this sounds, my description is a significant oversimplification of the actual procedures and all the decisions that the surgeon must make along the way as he or she uncovers the actual condition of the urogenital sinus and the relationship of vagina to urethra.

Masculinizing Surgeries

As complex as vaginoplasties are, building a functional penis is a nearly impossible task. According to the *Consensus Statement on Intersex Disorders and Their Management*: "The enormity of the undertaking and the complexity of phalloplasty must be considered during the initial counseling period. Care should be taken to avoid unrealistic expectations about penile reconstruction."

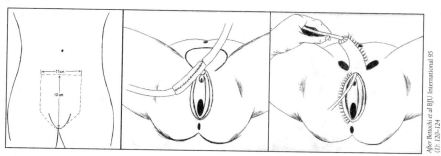

After Bettochi et al BJU International 95 (I): 120–124

The phallus is formed from a flap of anterior abdominal wall skin that is rolled around a catheter to form the new urethra and sutured to complete the new penis.

This is not especially encouraging for male expectations. But the few studies that have been done seem to bear this out. In one study of adults with gender dysphoria—meaning people in women's bodies who wished to be in men's bodies—penises were created using a flap of muscle from the abdominal wall immediately above the pubic area.

This flap was left attached at the base to the abdominal wall and sewn into a tube that contained a new urethra made from a tube of labial skin. Patients were then offered both penile and testicular prostheses—artificial elements for achieving erections and to fill an otherwise empty scrotum.

Of the eighty-five people treated and questioned between 1989 and 2000, most thought their constructed penis looked OK. But nearly everyone had problems including strictures (narrowing of the tubes) or fistulas (holes that form between two abutting tissues) involving the newly formed urethra. About 25 percent of them needed further surgery, sixteen had problems with urination, and three had some necrosis of the constructed phallus. Only 16 percent were capable of sexual intercourse without prostheses. However, sensation in the penis was possible in most, since the existing clitoris was incorporated into the new urethra.[17]

It is difficult to imagine the technical accomplishments involved in these surgeries. Working human flesh like clay to create, enhance, or define sex is truly one of the wonders of modern science.

Outcomes: Sex, Surgery, and Satisfaction

But does it work? Do these surgeries make for happier people? As previously mentioned, it depends upon whom you ask. Doctors, parents, and patients are all likely to provide different answers to those questions. The aftermath of surgery is often complex, and long-term studies of patient satisfaction are few and frequently contradictory.

Bruce, Brenda, and David: One Boy's Tragedy

During the summer of 1965—the same year Dr. Betty Suits Tibbs wrote her account of Lenore, the boy-made-girl (see chapter 1)—a pair of identical twin boys was born to Ron and Janet Reimer in Winnipeg, Manitoba—an old city built where the Red and Assiniboine Rivers come together in western Canada. In Winnipeg, the predominant faith is Roman Catholic, winters are harsh, and hard work is a way of life.

Ron was twenty years old in 1965; Janet was eighteen. Janet was pregnant when they married. Ron took a job at the slaughterhouse, and Janet settled into her role as soon-to-be-mother. The pregnancy unfolded without incident, and, though born a little prematurely, both boys were healthy and completely normal. The Reimers named the twins Bruce and Brian.

For the next seven months, Ron Reimer worked hard to earn enough money to provide for his new family. The boys did fine, and Janet was a good mother to them. But just as they were about to turn seven months old, both of the twins began to have trouble urinating. Each time one of them peed, he cried out in pain. Janet made certain it had nothing to do with their diapers, and then she examined both boys carefully. It appeared to her that the boys' foreskins were stuck closed. So she took both of them to see their pediatrician. When the doctor examined the twins, he explained it was nothing to worry about. Both boys suffered from phimosis, a condition that makes it difficult for boys to retract their foreskins. The doctor told the Reimers that circumcision was the cure, and after some discussion, Janet and Ron made arrangements for the boys to have the surgery. [1]

The operations took place at St. Boniface Hospital in Winnipeg, a teaching hospital with an excellent reputation. Since Ron Reimer had agreed to work a late shift at the slaughterhouse, the Reimers dropped the boys off at the hospital the night before the surgery.

No one imagined that anything could go wrong. But what happened at St. Boniface the next morning was horrendous. Experienced pediatricians nearly always performed the circumcisions at St. Boniface. But the morning of April 27, none was available. So the job fell to Dr. Jean-Marie Huot, a general practitioner. With the toss of a coin, the nurse lifted Bruce from his crib and prepped him for the first surgery.

Electrocautery was Dr. Huot's method of choice. Electrocautery is a technique that uses an electrically heated instrument to simultaneously remove the foreskin and seal the severed blood vessels.

When Dr. Huot first touched the electrocautery needle to Bruce's foreskin, nothing happened. The doctor assumed the instrument wasn't generating enough electricity to do the job, so he asked the nurse to turn up the power on the machine. She did, and Dr. Huot tried again. Still nothing.

Once again someone turned up the power to the electrocautery instrument. This time, when Dr. Huot brought the needle to Bruce's foreskin, there was "a sound just like a steak being seared." Someone immediately turned off the machine and sent for a urologist. The urologist installed a catheter so urine could flow from Bruce's bladder, but it was too late to save any of Bruce's burnt penis.

Once a full assessment of the situation revealed to everyone the seriousness of the gaffe, the hospital called the Reimers and told them "there had been a slight accident and they needed to see [the Reimers] right away."[2]

For the next ten months, the Reimers and Bruce's physician agonized over what they should do for the boy. In desperation, Janet finally contacted Dr. John Money at Johns Hopkins Hospital in Baltimore. Dr. Money was very clear about how much good they could do for Bruce at Johns Hopkins. Janet felt like someone was finally listening. Dr. Money was certain that the only path left to them was to make Bruce into Brenda. So, at twenty-one months of age, the doctors removed Bruce's testicles along with the very last remnants of his penis and begin his feminization—including crafting a vagina from a piece of intestine and beginning a lifelong course of female hormones. From then on, Brenda dressed as a girl and eventually, thanks to the hormones, developed breasts.

Throughout, Brenda's parents made every effort to raise her as a girl, which couldn't have been easy after nearly two years of treating Bruce as a boy, not to mention the constant presence of Brenda's identical twin brother, Brian.

Brenda was described as "having many tomboyish traits such as abundant physical energy, a high level of activity, and often being the dominant one in a girls' group." Still, the doctors were convinced that she was developing as a normal girl. "Her behavior is so normally that of an active little girl and so clearly different by contrast from the boyish ways of her twin brother, that it offers nothing to stimulate one's conjecture." As a result, these authors, especially Dr. John Money, concluded that "gender identity is sufficiently incompletely differentiated at birth to permit successful assignment of a genetic male as a girl."[3]

But when interviewed at age thirty, David (born Bruce, then Brenda) claimed that none of that was accurate. He said he had never felt comfortable as a girl. And when they were interviewed, his brother, father, and mother said they had sensed the same thing all along.[4]

Finally, at age fourteen, Brenda could stand it no longer.

"I['ve] suspected I was a boy since the second grade," she told her doctor.

After her declaration, Brenda had a mastectomy to remove her hormone-induced breasts, changed her name to David, and began testosterone therapy. At age fifteen and sixteen David underwent surgeries to reconstruct his ruined penis. And at age twenty-five he married a woman named Jane and adopted her children.[5]

David's experiences are among the most widely known and quoted of all the travails of children with assigned sexes. At the outset, perhaps because of the lack of hard facts, the broader implications of this story were hard to decipher. But even now, when most of the facts are apparent, it is still difficult to decide what the Bruce/Brenda/David story tells us about how the sexes of humans begin and our perceptions of those sexes.

One thing seems clear: surgery, hormones, dresses, and dolls weren't enough to convince this boy that he was a girl. David's story ended tragically. His marriage to Jane was rocky from the beginning. He often fell into prolonged dark moods that excluded nearly everyone. He felt inadequate and unable to perform his true duties as a husband. And during his depressive periods he often exploded in anger.

Then, in the spring of 2002, twelve years after David married Jane, his twin brother, Brian, committed suicide using an overdose of antidepressants. David fell into a deep despair. Shortly before Brian's death David had lost his job and had been swindled out of about sixty-five thousand dollars. Neither the loss of the money nor the job was critical financially (David had made a sizeable sum of money from selling his story), but the shame was considerable.

In May 2002 Jane suggested a trial separation. David left in a rage. Two days later the police called to tell Jane that David was dead.

Shortly after David had stormed out of the house, he'd returned to the garage, taken his shotgun, and neatly sawed off the barrel. Then he'd driven to a nearby grocery store parking lot and ended his life.[6] There the boy-made-girl-made-man emptied his head of shame, confusion, and pain.

What about the others, the people who have had sex-enhancing or -defining surgeries under better circumstances? Surprisingly, it wasn't until the early 2000s that the medical community got serious about asking those most directly affected by our decisions and surgeries how they felt about what we had done to them. Given the chance to speak, these people had some very interesting things to say.

When Surgery Is Essential, Does Sex Matter?

Cloacal exstrophy is not truly an intersex condition, but it is one in which sex reassignment through surgery is typically required, not optional. *Cloacal exstrophy* is the term physicians use to describe a rare type of abnormal fetal development in which the whole bottom portion of the abdominal wall fails to form properly. In these people, where their legs join, there is a hole. Through that hole, portions of the intestines and bladder often protrude outside of the body, and the genitalia, and sometimes the anus as well, do not form at all. The survival of these children depends on early and expert surgery. Because there is often a complete absence of genitalia of any sort, as part of that first surgery the doctors assign a sex.

The most commonly assigned sex for babies with cloacal exstrophy has been female, due most directly to the relative ease with which doctors could fashion a semifunctional vagina compared to an even barely functional penis. For genetic females, this approach seems generally satisfactory. But in genetic males with cloacal exstrophy, the testes form normally and produce normal amounts of male hormones before and after birth. So these boys' brains and bodies—like those of Bruce/Brenda/David—are steeped in male hormones for several months before a surgeon can remove their testes. Because of their peculiar situation, these people offer physicians an opportunity to assess the general relevance of the Bruce/Brenda/David incident.

In one long-term study conducted at the Oklahoma Health Sciences Center in 2003, eight of fourteen boys who had originally been assigned as females because of cloacal exstrophy declared themselves as males during the course of the study. Four of the boys chose reassignment spontaneously. The other four chose reassignment after "learning that they were born male." The authors of this study concluded that prenatal androgens were "a major biologic factor in the development of male sexual identity."[7] Some of these boys, like Bruce/Brenda/David, found that their minds could not accept the bodies that they'd been given. But that was true for only about half of the boys, making the likelihood of success a near coin toss. That made many people suspect that even the Bruce/Brenda/David incident was an anomaly, or just bad luck.

In support of this, another study, performed in 2004 at Columbia University in New York City, reached a similar conclusion. This study reviewed published accounts of 46,XY persons with penile agenesis (an extremely rare disorder in which the penis fails to develop at all), cloacal exstrophy, or penile ablation (similar to Bruce's experience). On the basis of their review of the literature, these authors concluded that the majority of 46,XY babies assigned as females did not choose to change their gender later in life, although there was a slight increase in gender reassignment as these girls reached adulthood. In the end, about half these people originally assigned as females chose to continue life as females, and about half chose sexual reassignment as adults.[8]

Clearly, some genetic boys arrived at adulthood uncomfortable with their assignment as girls. But a nearly equal number of other 46,XY individuals were comfortable as girls. Once again, the odds of success seemed about the same as choosing heads at the opening coin toss of a football game.

When Surgery Is Optional, Does Sex Matter?

The results of studies of people with cloacal exstrophy and related disorders—disorders that demand reconstructive surgery—don't seem to offer much insight into the best route to successful sex assignment. But what about other disorders of sex development—disorders in which surgery may be optional?

One of the first large-scale studies of intersex people who'd had optional surgery was carried out at Johns Hopkins in 2002. The studies involved seventy-five adults with a 46,XY karyotype. Most of them were diagnosed with "undermasculinized genitalia" before they were two years old.[9] The scientists performing these studies divided the participants into three groups. Group one contained eighteen people with normal-appearing female genitalia. These people looked like females because of either complete androgen insensitivity syndrome or complete gonadal dysgenesis. As mentioned previously, people with complete androgen insensitivity syndrome lack normal receptors for androgens including testosterone. And 46,XY people with complete gonadal dysgenesis have mutations in the SRY region or deletions of portions of the Y chromosome. All the people in group one were raised as females.

Group two was composed of eighteen people diagnosed with micropenis. These people's penises were less than three quarters of an inch long and with "male-typical urethral" openings (i.e., not enlarged clitorises with hypospadias). Micropenis can result from any of several different developmental alterations. Five of the people in group two had been assigned as females and thirteen as males.

Group three had thirty-nine people with ambiguous genitalia. These people had both a small phallus and perineoscrotal hypospadias—meaning that the phallus was small for a penis, and the urethra opened in a more typically female fashion, making it likely that the phallus was a large clitoris rather than a small penis. This happens in people with partial androgen insensitivity syndrome, partial gonadal dysgenesis, and some other genetic mutations. In group three, eighteen people had been assigned as females and twenty-one as males.

Basically, the purpose was to gather a group of people with different beginnings, surgeries, outcomes, upbringings, and expectations and ask them just how much they knew about their conditions and their chromosomes, how satisfied they were with what had been done to them as children, how their sex lives were, whether they felt they needed psychological help, and how much they enjoyed their lives.

When asked if they were generally satisfied with the surgeries performed on them as children and with the adult lives they now led, most responded yes, suggesting that 46,XY children raised as either males or females could lead satisfying and productive lives.

Second, among the thirty-nine 46,XY people with ambiguous genitalia, the twenty-one raised as men required a lot more genital surgeries than did the eighteen raised as women, which, considering the complexity of masculinizing surgeries, is not so surprising.

Third, physicians rated the cosmetic appearance of the genitals of those raised as men as much more "abnormal" than the genitals of those raised as women. Again—given the complexity of the undertaking—this is not too surprising.

Fourth, among the group of thirty-nine 46,XY people born with ambiguous genitalia, the majority (including nine of the twenty-one assigned as males and twelve of the eighteen assigned as females) were content with their body images. Among the others, only one in each sex group was totally dissatisfied with his or her body image. The rest (eleven of twenty-one raised as males and five of eighteen raised as females) were only somewhat dissatisfied with their body images. Among this same group of people, three of the twenty-one raised as males and three of eighteen raised as females were totally dissatisfied with their sexual function, six in each group were totally satisfied, and the rest somewhat dissatisfied.

And finally, 90 percent of these men and 83 percent of these women reported that they had had a recent sexual experience with a partner.

On the surface these results seem to say that, regardless of the underlying cause, the majority of 46,XY children who, for one reason or another, don't quite merit the status of normal male can be arbitrarily surgically and socially assigned as either males or females with an equal probability of success. That interpretation supports John Money's hypothesis that the sex of a child is amorphous, and it is only the fire of appearances and upbringing that forces the crystallization of male or female. Here at last, it seemed, there was solid evidence that surgery was, in fact, these children's salvation. And with the proper upbringing, a child could become anything his/her parents/physicians wanted him/her to be.

A second study made a similar but more narrowly focused assessment of seventy-two of the same seventy-five adult 46,XY patients included in the first study. The conclusions of this study were very similar to those described above. Thirty-two of these people were assigned male sex and forty female. The majority of these people said they were mostly satisfied with their assigned sex and had never felt unsure about their gender. Furthermore, the majority did not think that a third sex category should be created, did not think that their genitals looked abnormal (even though the majority of the men thought their penises were too small), were generally satisfied with their sexual function, did not think surgical correction of ambiguous genitalia should be postponed until adulthood, and believed that their own surgeries should not have waited until they were adults.

Again, the data suggested that most of these people could be either men or women and be happy.

Another study, carried out in England in 2003, addressed more directly whether 46,XY individuals might live happily as females.[10] As mentioned, complete androgen insensitivity syndrome causes 46,XY fetuses to develop female external genitalia. And because some testosterone is converted to estrogen at puberty, these people may experience normal breast development. But their complete androgen insensitivity prevents normal development of pubic and underarm hair. So most women with complete androgen insensitivity syndrome aren't aware of it until they reach puberty. Then, lack of menstruation leads them to a physical exam and the discovery of undescended testes. At that point, usually, a doctor removes their testes and begins estrogen therapy. The studies at Johns Hopkins had also examined this aspect of intersex and the satisfaction of XY females.

In both studies, all 46,XY persons with complete androgen insensitivity were satisfied with their assignments and rearing as females. In addition, in this study 46,XY women were similar to 46,XX women in all behavioral categories examined, had heterosexual erotic interests, and had married into heterosexual relationships as often as the control group.

Clearly, a Y chromosome is not enough to make a baby boy, nor does it make for abnormal or unhappy women.

But complete androgen insensitivity may be an exception. These children do not have ambiguous external genitalia. For the most part, they look, act, and feel like other girls. Nor does sex assignment require any early surgery. In fact, sex assignment occurs pretty much on its own. Usually it isn't until puberty that decisions have to be made, and these are different sorts of decisions from those made by children with ambiguous genitalia. So this syndrome and the people it affects fall, perhaps, into a special nook. Regardless, complete androgen insensitivity syndrome and the women it creates prove one thing—there is more to sex determination than chromosomes.

Sensation after Surgery: The Downside of Clitoral Modification

Early in the modern era of research on intersex, someone started, but never published, a rumor that some African tribeswomen who had undergone ritual clitorectomy continued to exhibit normal sexual responses. Therefore, according to this rumor, a clitoris was unnecessary for a woman's normal sexual function. Because of that, the solution to the problem of large clitorises was obvious: simply cut them off. So until the late 1970s and early 1980s, clitorectomy was the procedure of choice when sexual assignment included surgery.[11]

In the mid-1980s, though, some physicians finally questioned the prevailing wisdom about cutting off clitorises and suggested that that the clitoris might be "an erotically important sensory organ that was worth saving" during surgical sex reassignments.[12] It is surprising that it took this long for physicians to begin to question women about their surgeries and even more surprising that it took this long for the surgeons to appreciate the sensory importance of an intact clitoris, no matter how large.

Since that time, genitoplasties, often called clitoroplasties, have mostly replaced clitorectomies in sex-assignment surgery. The intent of clitoroplasty is to provide for a number of things including an acceptable cosmetic appearance (acceptable to both physician and patient) and (where possible) normal, satisfying intercourse in adolescence and adulthood. It has been generally assumed (especially by physicians, but also by parents) that the outcome of clitoroplasty is preferable to either clitorectomy or doing nothing.

For that reason, and based on patient satisfaction like that described in the Johns Hopkins studies, in several disorders of sex development, clitoroplasty and vaginoplasty remain the methods of choice. And clitoroplasty and vaginoplasty appear to offer great options to those born with disorders of sex development, their parents, and their physicians.

But again, these conclusions are mostly based on assumptions, not long-term studies.

Still, on the whole, these studies appear to suggest that many people who have undergone sex-clarifying or sex-changing surgeries as children have become contented adults, regardless of the sex assigned to them.

Or so it would seem.

Dr. Alice Domurat Dreger is associate professor of Clinical Medical Humanities and Bioethics at the Feinberg School of Medicine at Northwestern University in Chicago. She is a historian and bioethicist who specializes in what happens to people born with "socially challenging anatomies"—conjoined twins, dwarfism, people with severe craniofacial anomalies, and people with disorders of sex development—people whose appearance might make some of us feel uncomfortable. Dr. Dreger has worked extensively with people from the Intersex Society of North America as well as individual men and women with intersex. She has met many people who, as children, had surgeries intended to make their genitals appear more typical.

"I suspect there must be happy people out there, but I've never met one," Dr. Dreger says. "Well, literally, I've met one. I should have met more by now if, in fact, the numbers were anything like the reports say they are. I should be hearing from lots and lots more people who had surgery as children and are happy, and over twelve years of this work I've heard from one person who had a vaginoplasty as a child who is happy with it."

Maybe Dr. Dreger has simply been unlucky enough to meet only unhappy people. After all, the studies done at Johns Hopkins University Hospital involved the largest number of people of any published reports so far, and those investigators reported that a majority of people who had undergone surgical sex assignment as children were generally happy with the results of their surgeries and were well-adjusted adults. And there are thousands of people out there who have undergone genital surgeries to normalize their appearance. Because of her extensive involvement with intersexuals, it seems like Dr. Dreger should have met some of those happy people.

A similar discrepancy arose between the data from a study done in Great Britain and people in support groups. The research project was carried out at St. George's Hospital Medical School at the University of London and the Middlesex Hospital in London, and it involves the outcomes of interviews with eighteen women born with congenital adrenal hyperplasia. Congenital adrenal hyperplasia results in overproduction of several hormones, including androgens. That sometimes causes 46,XX baby girls to develop enlarged clitorises and labioscrotal folds that don't quite make either labia or a scrotum.

All of the women in this study were assigned and raised as females. Some (the authors don't specify exactly how many) had had genitoplasty to "feminize" their genitalia. The researchers' conclusion was that "women with congenital adrenal hyperplasia are psychologically well adjusted and do not show substantially increased psychiatric disorder or deficits of social adjustment compared with population data."[13]

However, Melissa Cull, a woman with congenital adrenal hyperplasia, sees things differently.

"Although I am glad that the small cohort of women in the paper had good outcomes, support groups tend to hear a somewhat different story.[14] Perhaps only people who are dissatisfied with their treatments come to support groups—though going by the rarity of the condition and the comparative number of members it shows that many are, not surprisingly, dissatisfied—but we do get people who are satisfied as well."

Ms. Cull goes on to describe how many support group members report major depression and stress, particularly with relationships, sexual difficulties after surgery, and weight gain due to steroids.

"Non-disclosure [i.e., doctors not revealing full medical histories to their patients], shame, secrecy, and stigmas attached to having ambiguous genitalia, and intersex condition, and surgery to 'normalise' all place a heavy toll on woman's psychological well being," says Cull. "Many women with congenital adrenal hyperplasia avoid social situations, frightened that people will find out they have a rare, misunderstood condition."

Like Dr. Dreger, Ms. Cull has met a few satisfied customers, but too few if one believes the published data.

While Dr. Dreger's and Ms. Cull's evidence is anecdotal and not the result of large systematic studies, their stories are based on broad personal knowledge of individuals with intersex and the problems they face. But there are two studies out there that support both women's personal experiences.

Concern About Clitoroplasty

In 2004, a group of scientists working at University College of London Hospitals questioned and examined six women who had congenital adrenal hyperplasia and had undergone clitoroplasties as children, though in some patients it appeared that clitoral reductions may have more nearly approximated clitorectomies. The researchers found that all of the women had abnormal sensation in their clitorises, including abnormal responses to heat, cold, and vibration, and all engaged in intercourse less often than the control group.[15]

For this group, clitoral surgery clearly altered clitoral function. It is, of course, impossible to say how their sexual function might have differed without surgery.

This study was then expanded to include thirty-nine adult women. Twenty-eight of them were currently sexually active, but all of them reported difficulties with sexual intercourse. Nearly 80 percent of those who had had clitoral surgery reported clitoral insensitivity and inability to achieve orgasm. These researchers concluded that "Sexual function could be compromised by clitoral surgery," and that further debate should take place on the ethics and use of clitoroplasty in intersex patients.[16]

Clitoroplasty may seem to be a relatively straightforward procedure, but for several reasons the outcomes are difficult to predict. In all patients, the appearance of the genitalia will change with age, so a patient's satisfaction with the appearance of her clitoris is likely to change as well. At puberty, the appearance of pubic hair and fat deposits in the labia can dramatically change the look of the genitalia. And as a girl's body changes size and shape, an enlarged clitoris may become more or less noticeable. So in spite of its popularity as a "cure" for several types of intersex children, no one can really predict the outcome of clitoroplasty on sexual function, cosmetic satisfaction, or sensation.[17]

Still, overall, the number of clitoroplasty patients interviewed remains rather small, so it is difficult to draw general conclusions from these few studies and personal observations.

On the other hand, the Johns Hopkins studies were larger and more systematic. That does add to their credibility, but these studies are not without problems of their own.

Concern About Studies of Sex Satisfaction

Joel Frader is professor of Pediatrics and professor of Medical Humanities and Bioethics at Northwestern University's Feinberg School of Medicine. He is also Head of General Academic Pediatrics. He studies intersex. According to Dr. Frader, "The trouble with those [Johns Hopkins] papers is that they surveyed, usually, by phone—not necessarily the best way to do things. And, they surveyed their own patients. For me, the issue there is the bias involved."

Dr. Frader explains, "If I'm taking care of a patient, and I want to know whether that patient is happy with my treatment of him or her, for me to call the patient up and say, 'We've been looking over your chart. Would you tell me whether you think things have gone well?' There's a real problem having to do with the relationship between the treating team and the patient, in which patients don't want to disappoint their doctors, especially if the patients are still going to [the doctors] for medical care."

And it isn't just the Johns Hopkins studies that seemed flawed. Dr. Frader believes that there are just no good data out there, period—no study that believably and decisively answers questions about how kids who underwent genital normalization surgery feel about it as adults.

While the Johns Hopkins studies involved a mix of people with partial or complete androgen insensitivity syndromes, complete or partial gonadal dysgenesis, and other unidentified causes of ambiguous genitalia, at least two other studies have examined outcomes for 46,XY people born with cloacal exstrophy—a developmental disorder that often demands reconstructive surgery simply to ensure the child's survival. As mentioned previously, about half of these patients chose reassignment to male sex as adults; in one study, nineteen of twenty patients suffered from one or more anxiety disorders.[18] This would seem to contradict the studies done at Johns Hopkins.

But, according to Dr. Frader, "Hospitals like mine have exstrophy patients—all academic children's hospitals do—but those children are a small minority of the total population of children with disorders of sex development. So, I don't know what one can make of that."

The other potential problem with these studies is the nature of the condition itself. Cloacal exstrophy is a devastating and life-threatening disorder. Simply surviving it is a major accomplishment, and there are often other major, non-sex-related disabilities associated with the disorder. If, in addition to survival, a child can urinate nearly like other children, look a little like either sex, and lead an approximately normal life, parents and children may find little room for complaint about even major disappointments and are, thus, less likely to consider sex reassignment. And then there are the repeated surgeries required and the sheer difficulty of creating a penis-like phallus. So the motives of those who did not choose further surgeries may be many, only one of which could be their level of satisfaction with their original surgeries.

Alice Dreger shares Dr. Frader's concerns about the studies that have been done. She points out that, of the 183 patients originally identified as candidates for the Johns Hopkins studies, only seventy-five responded and were deemed competent to participate. That means that more than two-thirds of the subjects eligible for this study chose not to participate, could not be found, or were disqualified for one reason or another.

It is impossible to determine what sort of inadvertent and unavoidable selection may have taken place just in identifying and contacting eligible patients. It is equally impossible to read the minds of those patients who chose not to respond. But if even half of those who chose not to participate in the study did so because they were unhappy with the outcomes of their surgeries, that would change the interpretation of the study completely.

Dr. Dreger is equally concerned about results gathered by physicians interviewing their own patients, noting, "When you're asking about something sensitive, you actually want to engage a third-party professional—a psychologist or a sociologist—to ask, in in-depth, interactive interviews, what happens, and [those at Johns Hopkins] haven't done that."

And there are others who have concerns with these studies. Most quality-of-life studies performed by physicians have identified (often as a significant majority) a larger percentage of people who are content with their lives and generally satisfied with the results of their childhood sex-assignment surgeries than those who are not satisfied. But as doctors Ursula Kuhnle and Wolfgang Krahl from the Children's Hospital at the University of Munich and the Psychiatric Hospital in Kaufbeuren, Germany, point out, there are alternative interpretations of those studies, including those performed by Drs. Kuhnle and Krahl.[19]

In nearly every one of the published studies, the investigators concluded that, because the patients had no significant complaints about their surgeries and were functioning as contributing adults, the physicians' treatment and management decisions had been the right ones. But Kuhnle and Krahl object to this assumption: "We now believe that this is an unwarranted conclusion, and that the only inference to be drawn from these studies is that the majority of these patients are well adjusted and can somehow live with a handicap. These studies do not answer the question of whether there are other and/or better options for a patient's life."

In other words, perhaps these people are more or less content with their lot in life because they, like many other human beings, are remarkably resilient and have learned to live well with a major problem. Nothing about the way in which any of these studies have been conducted even begins to approach the question of whether these people would be happier if their physicians and parents had chosen differently, or done nothing at all.

8

ALTERNATIVES: OTHER CULTURES, OTHER SEXES

From atop his milk-white bull, the four-armed creator and destroyer Nataraja watches over the world. He is called Bhairava as well, the fearsome god who cut off one of Brahma's five heads and carries it with him while he guards the pieces of his wife's corpse. And he is Lord Agni, the seven-armed, two-faced god of fire. He is the cloud-colored, four-armed Vishnu, the all-pervading essence of all beings. He is Ardhanarishwar, the curious blend of god and goddess, from whom came the reproductive prowess of all living things. And he is the *lingam*, or mythical phallus from which all creatures sprang—a symbol older than all of Hinduism. In myriad incarnations as different from one another as night from day, he is always Shiva the Divine.

Shiva as Ardhanarishwar.

When asked to create the world, Shiva undertook the task with such diligence and reflection that, for millennia, nothing happened. So the power of creation was given to Brahma, who quickly moved to

143

squeeze a universe from the darkness. When Shiva finally gathered everything he needed for creation, he was stunned and angry to find the universe already in place. In his rage, the god tore off his own phallus, screaming "there is no longer any use for this," and hurled the severed member to Earth. Once separated from the god, and no longer a tool for his individual fertility, Shiva's phallus became the fount of universal fertility.[1]

Brahma, meanwhile, discovers that his great creation, the universe, has become stagnant and sterile. He realizes that because every created manifestation of Shiva is male, copulation and reproduction are impossible. So Brahma begins to contemplate the forms of Shiva along with an image of the goddess Shakti (though some say it might have been Parvati or even the Dark Mother, Kali). Shiva senses Brahma's thoughts and appears to Brahma as the god Ardhanarishwar—part Shiva, part Shakti—something between or beyond man and woman. Brahma is completely taken by Ardhanarishwar and the fertility of Shakti. He begs her to allow him to bestow her fertility upon his creations. Willingly, Ardhanarishwar gives to Brahma that which he most needs, and Brahma's creations copulate and multiply.

Like Hinduism, many other religious traditions speak of deities and humans who are neither men nor women, including the androgyny of the Judeo-Christian Adam. But those are very old stories that have passed through many hands. Much may have changed or been lost in translation. A better way to test the foundations of the two-sex mythology would be to look at the peoples of our modern world and see if such beliefs are universal among human societies. Do people raised with different worldviews see the sexes differently? The answer is an emphatic yes.

The *Hijras* of India

Among those who honor Shiva's destruction of his own phallus and his manifestation as Ardhanarishwar the intersex god are a group of Indian Hindus called the *hijras*. Literally, *hijra* means "man-minus-man."[2] Sociologically, it means much more than that. *Hijras* are not men, but neither are they women, and in their own country they were not always reviled, as some might expect. Rather, many have attained an exalted social status.

In India, one of the greatest social events is the birth of a son. When a daughter marries, the girl must bring with her a large dowry or the marriage won't happen. A string of baby girls can bring a family to financial ruin as quickly as a robber, while a string of sons can add great wealth. Because of that, a baby boy is considered a wonderful gift, and at their birth, the *hijras* perform one of their most important functions. The *hijras* sing traditional and popular songs, imitate the process and the pains of pregnancy, and bless the parents and the baby. Then, at a certain point in the ceremony, one of the *hijras* closely examines the newborn child to confirm that it is indeed a boy. If the *hijra* proclaims the child a boy, then the celebration proceeds and intensifies. But if the *hijra* sees that the child is intersex, all of the *hijras* immediately claim that the child is one of them and belongs to their community. And it is likely that the *hijra's* words will hold true. Many intersex Indian children do become *hijras*. But intersex children are not the only children who become *hijras*.

Hijras are not exactly men because, for one reason or another, they are incapable of performing the male role in sexual intercourse. And they are not strictly homosexuals, though some do have receptive sex with other men. But they do not, because of course they cannot, have sex with one another.

Hijras are not exactly women, even though they do dress, wear their hair, and act somewhat like women. But unlike most Indian women, *hijras* may be aggressively sexual, dress garishly, dance in public, and curse offensively. And many *hijras* never have sex with men.[3]

Baby girls cannot become *hijras*. It seems that the female sex role is more fixed in Indian tradition than that of men. For that reason, some Indians and especially outsiders think of *hijras* as defective men. But they are much more than that. And even though all *hijras* will claim they have been as they are from birth, not all were born intersex.

Hijras are not simply intersex or impotent males. They receive their calling from a Bahuchara Mata, the Mother Goddess, who is also honored by transvestites. And that calling can come to physically normal boys as well as to intersex boys. If any child ignores the call, it is said he will be impotent in his next seven lives—an ignominious fate.

Boys who receive the calling and are not intersex must undergo a dangerous surgery before they can call themselves true *hijras*. Interestingly, *hijras* call that surgery "the operation," in English. A special *hijra* called a midwife performs the operation, which replicates many of the events of childbirth. With two quick cuts, the midwife removes both penis and testicles. Because the blood is part of the male principle, it is allowed to flow freely to rid the body of as much of the male as possible. A small hole is created for the urethra, and traditional rather than modern treatments are used to induce healing. After the surgery, others treat the new *hijra* much like they would treat a woman who has recently given birth. After healing, the new *hijra* dons a bridal costume and parades through the streets in a procession with other *hijras*. Once a boy, now something else, the person walks into a new life, something beyond man or woman.

Men Not-Men Among the Natives of North America

The *hijra* of India are not the only people who fall outside of our easy sexual dualism. The same is true for some North American natives. The recently arrived Europeans called these people *berdache*, a bastardization of an Arabic word meaning "prostitute." These Native Americans are not prostitutes, so the word *berdache* is a poor fit and in many ways insulting. But because of language differences among native peoples and confusion among nonnative peoples there is no other title that people agree on.

A word that has gained some popularity is the name "two-spirit." But there are problems with this term as well. Coined around 1999, the word was actually invented by urban gay and lesbian Native Americans. The term has sometimes been co-opted to include the *berdache*. That seems inappropriate for at least two reasons. First, there is some evidence that gay and lesbian native Americans coined the word to distance themselves from the *berdache*. [4] Secondly, the *berdache* are neither gay nor lesbian. That doesn't seem to fit with the way the *berdache* see themselves.

Berdache have also been described as homosexuals because they do at times have receptive sex with men. But the term *homosexual* has no meaning within the culture of the *berdache*. Native American men having sex with *berdache* do not consider this a homosexual act, because *berdache* are not considered men.

Europeans also called the *berdache* transvestites and hermaphrodites, and neither of those terms is accurate either. The problem with English and French (not to mention all the other European languages) is that those languages were born of cultures that saw the world differently from indigenous North Americans. Our words cannot easily hold this other world.

The names for these people vary from tribe to tribe, as do their cultural roles and their lives. Among the Navajo, these not-men are called *nádleeh*, and the Mohave call them *alyha*. In other tribes, they are absent altogether. And even though there are dramatic differences in the behavior and the roles of the *berdache* between tribes, there are many similarities. All cross-dress to some extent, all take up the occupations of the other sex, all have sex with men (but not exclusively), all use special language, and they all are, or once were, recognized as having spiritual power.

Most often, the boys who will become not-men manifest themselves in childhood by their fascination with the things and the ways of women. Interestingly, in most tribes where there are *berdache*, there are also women who are not-women. And like the boys, these women usually identify themselves in childhood by their fascination with weapons and horses. So the Navajo recognize four sex/gender variants: men, women, female-bodied *nádleeh*, and male-bodied *nádleeh*. Sex between male *nádleeh* and men or between female *nádleeh* and women is accepted by the Navajo as a normal form of behavior. However, there are strong taboos against sex between two men, two women, two female *nádleeh*, or two male *nádleeh*. Not-men are in fact not men, and not-women are something beyond other females.

The social roles of these people can be extremely complex. One remarkable example of this happens with the *alyha* of the Mojave Indians. The following description relies heavily on information in the book *Gender Diversity* by Serena Nanda.[5]

Alyha, though they have penises and testicles, do not see themselves as men. And even during sex, they ignore their erect penises and expect their partners to do the same. As *alyha*s reach adulthood, they actively seek husbands. Once an *alyha* finds and marries a man, she begins to simulate menstruation. She does this by cutting herself between her legs with a sharp stick or stone. Her husband and her friends treat the *alyha* just as they treat any other Mojave girl who has entered puberty and menstruated for the first time.

As the marriage continues, about once a month the *alyha* cuts herself and bleeds. But after a time without children, some husbands of *alyha*s threaten them with divorce, because children are important to the Mojave. When that happens, the *alyha* stops cutting herself and bleeding, and she begins to observe all of the tribal taboos that surround pregnancy. Then she publicly announces her pregnancy, something most Mojave women never do. For the next months, the *alyha* stuffs increasingly larger rags into her skirt. Then, at about nine months, she begins to drink a tea that will constipate her. That goes on for a few days until her stomach pains become so severe they are—just like labor pains—apparent to everyone, and she can no longer stand them. Then she goes into the trees and bushes, digs a hole and, in the same position as a Mojave woman giving birth, the *alyha* defecates into the hole. She and others treat her feces as they would a stillborn child and bury them. The *alyha* and her husband cut their hair and descend into a period of deep mourning, just as Mojave couples who have lost children do. And the *alyha* has saved her relationship with her husband.

For the sake of convenience, I have used the pronoun *she* throughout this description, but clearly, *alyha*s are not exactly women.

And though *berdache* often took up the jobs of women in their societies, their roles varied. They were not, as some trappers and others suggested, cowards or failed men. One fascinating example is the Crow Indian Finds Them and Kills Them. Finds Them and Kills Them's dress was striking. In one picture he wears the headband of a warrior and the dress of a woman. In one hand he holds a knife, in the other a shawl. A purse fits tightly into a beaded belt, and what looks like clam or oyster shells are stitched all over the upper half of the dress. In the role of *berdache*, Finds Them and Kills Them carried on the traditions of women. In war, Finds Them and Kills Them was as fierce as any Crow warrior and was known far and wide for his/her great bravery, especially among the Lakota Sioux, the Crow's greatest enemies.

Thinking Inside the Box

It isn't only in India and among the native North Americans that such people walk. "Not-men" live among the Muslims of Oman, where they are called *xaniths*[6]; among the people of Brazil, they are known as the *tranvesti*, *viado*, and *bichas*; among the Hawaiians and Tahitians they are the *māhū*; in Thailand they are called the *kathoey*; and in the Philippines they are the *bakla*. It is likely that there are many other peoples in other places who similarly fall outside of our definitions of men and women.

The simplest explanations for these people might be that they are simply men or women with significant identity issues, or that they are simply homosexuals. What would we have found if we had just lifted the dress of Finds Them and Kills Them and peered between his legs? Likely we would have found ourselves dead. But even if we had survived, whatever we found would have been of no significance. We've already seen that genitals can be a poor measure of a person's sex, and so can any other criterion we might choose—chromosomes or hormones, and on and on. Even if we wished to explain all the ambiguity away as some simple sort of genetic mutation, we have to ask ourselves, why does that mutation keep popping up? Why does it appear in so many societies? Generally, these people aren't reproductive. Biologically, that should be a death knell for their genes. But the people and their genes don't go away. Perhaps all of us are essential in ways we have not yet imagined.

At least some of those who know the *nádleeh*, *hijra*, and the *alyha*, those who see them with other eyes, have great respect for them and their spiritual powers. Those who know them through other eyes don't see what we see, don't feel the need to trivialize them somehow and squeeze the spread of human beings in the vise of our words. They don't feel the need to question what the universe has given to them.

Or at least once they didn't. When Europeans first began to explore the Americas, only the northeastern, north-central, and east-central tribes lacked *berdache*. Now the *berdache* are relatively few. The metastasis of white religion and white values might have killed the *berdache*, or it might have been simple white ridicule of the *berdache* that doomed them. The murderous power of that ridicule is apparent today among the Pima and the Papago Indians of the American Southwest. *Berdache* still survive in these tribes, but they are condemned and mocked by many of their people.[7]

What About Their Parents?

No matter who we become, much of what set us upon our paths was out of our hands. Long before we even realized that choices had to be made, our parents were making those choices for us. And for all their lives, the world whispered in our parents' ears about what was wrong and what was right, what was cute and what was travesty. It is impossible to imagine that the cultures of our parents didn't limit the options open to us, doesn't still limit the options open to us, especially to those of us who have unexpected, maybe even inconvenient, needs.

Mojave boys normally begin to show interest in the activities of men— hunting, riding horses, making bows and arrows, interest in girls—by about age ten or eleven, before the tribal puberty ceremony. A boy who skipped all of this and instead picked up dolls and imitated the domestic work of women or forced himself into women's gambling games, a boy who chose a bark skirt over a breechcloth, was a potential *alyha*.

The parents' response to this behavior was reportedly ambivalence. Ambivalence is not the same as indifference, though we sometimes use it as though it was. Ambivalence means the simultaneous presence of strong opposing emotions—love and hate, fear and pride, promise and failure.

In the beginning the boy's parents try hard to force him into the activities of men. But when the child sticks fast to his mother and the ways of women, his family gives in resignedly and prepares for the "transvestite ceremony." But they tell the child nothing about this. At the ceremony, two women lead the boy into a circle of women who begin to sing "the transvestite songs." If the boy dances with the women he is an *alyha*. When that happens, the women take the boy to the river, wash away his past, and give him girls' clothes to wear. Afterward the *alyha* chooses a girl's name by which he will be known from then on.

Watching their child transform into an *alyha* pleases no parents. They know it will change their child's life forever and in some ways forever separates parents from their child. Still, in the end they accept the reality of the situation and move on. Beyond the transvestite ceremony they make no further effort to force their child into the mold of a boy. From then on, everyone agrees things will be different.

How Our Past Becomes Our Future

There has never been a large-scale study that investigated the effects of our cultural heritage on the decisions we make when one of our children doesn't fit with what we've seen on TV, in magazines and motion pictures, in showers and schools. Nor has there been a broad study of what might happen if we made no decision at all. But there has been one rather small study that produced some provocative results.[8]

Ursula Kuhnle and Wolfgang Krahl carried out this study at Malaysia's largest children's hospital at the University of Malaysia. The people of Malaysia mostly comprise three ethnic groups; about 60 percent are Malays, about 30 percent are Chinese, and about 10 percent are East Indians. The Malays are Muslims, the Chinese are Buddhists or Taoists, and the Indians are Hindus. The social roles of men and women, as well as the structure of the nuclear family, differ significantly among these peoples. The purpose of this study was to try to determine if these differences affected the way parents responded to the birth of an intersex child and the choices that birth necessitated.

Muslim women in Malaysia have roles in religious and public life similar to those of Christian and Jewish women, including the right to control their own money and, in the event of the death of their husband or divorce, to inherit their own money. The roles of women in Chinese and Indian societies are much more restricted. As we have seen, the birth of a girl is often perceived as a burden by an Indian family. And in China, men are both the head of their families and the ones who, through their work and fortune, can ensure that families will persist.

The authors' roles at the hospital in Malaysia were to attend to the births and management of intersex children. During their time at the hospital, they found it "was never difficult to convince a Muslim family to assign a severely virilized girl or undervirilized boy to the female gender. This was not the case for Chinese or Indian parents, who on several occasions took off with their ambiguously born child when female sex assignment (or reassignment) was suggested."[9] Just the idea of another daughter was enough to make the parents bolt.

The authors point out, though, that, while it was easier to work with Muslim parents, these people still took a diagnosis of intersex very seriously. And a few of the Muslim parents chose not to follow the doctors' recommendations at all.

In one case, a child by the name of Fatimah was born with congenital adrenal hyperplasia. At her birth, the doctors assigned Fatimah as a girl. For that assignment to work, any child needs to strictly follow a course of medications to control her overactive adrenal glands.

Fatimah's parents didn't think the medicines were so very important. So, at puberty, Fatimah further virilized. Her facial hair, body shape, and voice all became more like those of a boy. In addition, she preferred the company of boys and was accepted among them as a peer, "not as a tomboy who likes to play with boys."

Often parents like Fatimah's, who failed to comply with the doctors' instructions, saw the changes that occurred at puberty as proof that the doctors had got it all wrong—an affirmation of the parents' suspicions and a reason to go forward as they had originally planned.

Fatimah took up the normal roles of a boy in her religious practice, and at the appropriate age she was circumcised. Her acceptance as a boy was complete. The only reason the authors ever saw her again was because she later applied for an identity card, and the official in charge questioned the name Fatimah for a male and requested the doctors' input.

But things don't always turn out like that. In a second case, a child named Diana was referred to the authors at the hospital in Malaysia because of her mother's concern about the child's abnormal genitalia. The primary care physician had agreed with the mother that Diana's genitals were abnormal, but he said he couldn't tell if she should be a boy or a girl. So he suggested that Diana and her mother return when Diana was twelve or so years old, and they would go from there. Diana and her parents were of Chinese ancestry.

At home, Diana's behavior was much more like that of a Chinese boy than a girl. She wore shorts and T-shirts and played in many of the games with the boys. When the authors first saw Diana (at age twelve) she seemed built like and carried herself more like a boy than a girl. When the doctors examined Diana they found "a phallus of about 5 cm [approximately 2 inches] in length with significant erectile tissue, severe hypospadias [the opening of the urethra was at the base of the phallus, not at the tip] with a single penoscrotal opening. . . ." Using genetic tests, the authors ruled out things like 5-alpha reductase deficiency, but the ultimate cause of her intersex wasn't entirely clear. Regardless, Diana definitely had severely unusual genitalia, and partial gonadal dysgenesis seemed the likely culprit.

At her first visit, Diana talked with a psychiatrist. She claimed she was a girl, and asked for the surgery that would make her look more like one. However, after a little reflection, Diana changed her mind, as did both her parents. Diana would have been their second girl, and two girls would have been at least one too many for their way of thinking. Everyone, including all of the doctors, agreed Diana should be a boy. Appropriate surgery was initiated to, among other things, repair the hypospadias. But as a result of the surgery Diana's urethra developed strictures and fistulas (in this case between the interior of Diana's urethra and the surrounding tissues). Diana had to return to the hospital multiple times for further therapy and treatment. But the family lived far away from the hospital, making frequent visits difficult, so it was impossible to provide adequate psychological care. Previously an excellent student, during this time Diana's school grades plummeted, and eventually he simply disappeared.

Only a very few cultures are built on mythologies wise enough and strong enough to comfortably accept intersex children. Most of us living in the Americas and Europe have no such traditions. Or maybe I should say that most of us have only a single tradition, a myth handed down to us from Realdo Columbus, who imagined he had discovered the clitoris, and from the other great minds of the Enlightenment. That myth tells us that among humans there are two—and only two—polar, opposite sexes. Faced with a child who lives completely beyond what most of us ever imagined possible, that simple myth is a poor guide.

9

CHOICES: EXPLORING THE OPTIONS

In light of the poverty of our single myth, the nature of our world, the skills of our surgeons, and the needs of our children, what can we do? How can we wrest the best from our lackluster and often confusing options? The answer, of course, is that no one knows for sure. Every child is different. But people who have spent their lives wrestling with these questions agree on some things.

Dr. Alice Dreger has devoted much of her academic life to the study of both disorders of sexual development and, even more important, the people involved—men and women—not just syndromes. She has observed:

> The thing that people with intersex suffer from most is shame, it's not surgery. The surgeries are motivated by shame. So I think the bigger issue is people getting the message that [people with intersex] are not human, that they are not acceptable, that they are not loveable. That's a much bigger issue for everybody than the surgery is really. Because, what people who are anatomically different need the most is the message that they're human and acceptable and loveable.[1]

Regardless of what decisions physicians and parents may face, it seems Dr. Dreger's words should be foremost in their minds. Though these words guarantee no particular outcome or degree of happiness, they at least ensure that all other decisions must support this basic tenet of humanity, acceptance, and love.

Beyond that, things become somewhat cloudy. For example, how do we distinguish optimum outcomes from survivable outcomes? Well, one obvious way would be to include control studies that evaluate the satisfaction of people who did not have childhood surgeries as part of their sex assignment. According to Dr. Dreger, no such study has ever been done.

"Historically, what we know about people who grew up without corrective surgery [is that they seem] to have done generally OK, as far as we can tell.

"Instinctively, it seems that it should be true [that people with ambiguous genitalia would have greater psychosocial difficulties than the rest of us], but it also seems instinctive that no one would want to live as a conjoined twin."

And that is not generally true. In fact, Dr. Dreger has found that many conjoined twins do not wish to be separated, and many people with intersex do not wish to be made more like other men or women. "Just because something is statistically abnormal doesn't mean that it's bad. A lot of people who have these different conditions, not just intersex, say the same thing, which is: 'I know this is abnormal compared to the general population, but it's normal for me.'"

Lisa May Stevens bears out Dr. Dreger's beliefs. Lisa May has told me more than once that she no longer wishes she were someone else. I have met with Lisa twice since she first wrote to me, and both times it was clear that she revels in her chimerism, her hermaphroditism, and the novelty of it all. In fact, though many of the people I spoke with wished for more information, either earlier in their lives or even now, not one of them ever told me they wished they were different, more like others. No one.

Imagine that. Imagine, for just a moment, that some of these people have no wish to be exactly like others. Imagine happiness in singularity.

Dr. Dreger has "time and again suggested to the physicians doing these studies that they should [also] study that population of people who didn't have surgery. But they told me, 'Well that's not for us to do, because they're not our patients.'"[2]

It looks like it may be some time before we have those data.

So what do we do in the meantime? If you ask this question of a surgeon, he or she is likely to suggest surgery. But that is not the only course, and we know that some people untouched by a knife can achieve happiness, or something a lot like it.

With the exception of a few conditions such as cloacal exstrophy, choosing to postpone surgery does not seriously limit a child's later options. A potential advantage is that the child will have his or her genital and gonadal tissue intact and can decide if he or she wishes to change that. That would obviously be of major importance if the gender assignment made at birth turns out to be the wrong one for this child.

The process by which each of us comes to identify with one sex or the other is complicated. And which sex a child relates to—that is, his or her gender identity—results from the interaction of many different factors, including genetics, prenatal environment (including whether the child develops next to another boy or girl fetus), pre- and postnatal hormones, and childhood psychosocial and environmental factors.

According to a report in the *Archives of Childhood Diseases*, gender-identity selection is a process that lasts for at least the first five years of a child's life. At one year, most children can easily distinguish men from women, and a few children already show some preference for more masculine or more feminine toys. But it isn't until they are two or three years old that children regularly and correctly identify their own sex and the sex of others. The exact significance of that isn't altogether clear since it may simply be that the child cannot articulate their gender identity in a way that is meaningful to adults until the child is two or three years old.[3]

By age three, most children relate strongly with one sex and have a definite sense of being a boy or a girl. Interestingly, the genitalia of children and adults don't figure preeminently into a child's decisions about someone's sex. Instead, and maybe not so surprisingly, it is the more easily visible features that aid in a child's decision about another's sex—things like clothing, hair length, and size. But, even if a child can see another child or adult's genitalia, what they see is relatively unimportant to that child's decisions about that person's sex until the child is around age eight.

And it is at about age five that children come to believe that sex is unchanging with time. This is the point at which many people think that a child's gender identity becomes fully established and fixed. Then all the child's energy seems to focus on adopting behaviors consistent with that sex.[4]

That means that for the first several years of a child's life, some aspects of sex remain molten, and they only solidify after months of hormones, contemplation, indoctrination, and socialization. And, perhaps most important, it appears that children's genitalia play a relatively small part in this process, and that, though the data are still few, the studies that do exist suggest that unusual genitalia do not interfere with a child's development of a gender identity.

Research that followed the development of babies with ambiguous genitalia showed that the condition of the children's genitalia had little effect on the development of gender identity. 46,XY children with feminized genitalia developed male gender identity following almost the same schedule as boys or girls with more traditional genitalia.[5] Clearly, under the right circumstances, ambiguous genitalia are no impediment to becoming a boy or a girl. It seems that left to his or her own devices, an intersex child may find his or her own gender identity as easily as any other child.

But Dr. Frader and Dr. Dreger both believe that gender assignment at birth is critical to this process. According to Dr. Dreger:

> We don't have any social system for accepting as human somebody who doesn't have a gender assignment as male or female as a child. There are some adults who manage and struggle with having no gender, or having a third gender, or having two genders, but there is no social system at all for a child like that. I think it would be incredibly cruel and incredibly damaging to try to raise a child without a gender.

But clearly gender identity development does not depend on surgery. According to Dr. Frader, there is no convincing reason to do early surgery, and waiting until the child is old enough to have meaningful input is often a better way to go. He also argues that from the very start it is essential to have a multidisciplinary team—including pediatricians, surgeons, endocrinologists, ethicists, psychologists, geneticists, genetic counselors, and social workers—involved in the process. That is the best way to optimize the likelihood that the child will develop into a healthy and perhaps happy adult.

But even with all of that expertise, people, especially parents, must realize that the initial gender assignment may still be wrong. Dreger says, "I always say to clinicians: 'You've got to get parents to realize that for many types of intersex conditions, the odds of us getting the gender assignment wrong at birth is higher than that for the general population." In other words, "For most people with intersex, they'll never change their gender assignment. . . . But for some intersex conditions, like 5-alpha reductase deficiency, and like intersex conditions involving ovotestes, there's a higher rate of what I'll call gender unpredictability—that people either change gender later or don't quite feel the gender they were assigned."[6]

That would seem to be among the best arguments for not performing irreversible surgeries while the child is still too young to have any role in the decision-making process—that and Alice Dreger and Melissa Cull's observations that a lot of people out there are unhappy about what was done to them surgically as children.

Are Our Attitudes and Our Medicine Changing?

Yes and no.

Dr. Frader observes, "I think it's highly unlikely there's been a complete revolution in practice. I think there are still a lot of places that have pediatric endocrine leadership . . . who are still very much in the mode of, 'We make paternalistic decisions. Here's what *we* think should happen,' including counseling families to maintain secrets. But, I think that's becoming less and less acceptable."

In fact, it seems that many physicians' thoughts about management of intersex children have undergone major changes very much like those suggested by Dr. Frader—more parental involvement and input from pediatricians, surgeons, endocrinologists, ethicists, psychologists, geneticists, genetic counselors, and social workers. And changes in our vocabulary.

Curiously, though, it isn't clear that physicians' changes in attitude are translating to changes in practice. According to Dr. Dreger, that sort of change will take time. The general public, though, is a bit faster to respond. Some parents and children may yet find themselves working with physicians who insist on surgery and who may choose not to involve too many others in their decision-making process. General changes in medical practice can be glacial in their progress.

According to Dr. Dreger, "The lay audiences are much more receptive to this message. I think there are a couple reasons for that. One is what I call the *Will and Grace* effect, and I really do mean *Will and Grace*. That TV show and similar ones that have put gay and lesbian life out in the public eye [have] made an enormous difference toward people accepting people who are gender nonconforming. People with intersex are often gender conforming, but they're sex nonconforming. So it all gets sort of mashed together. So the effect of the gay and lesbian population rights movement has made a huge, huge difference in terms of humanizing people with intersex."

She also attributes some significant changes in attitude to Jeffrey Eugenides's book *Middlesex*, a remarkable novel that describes the difficulties a boy experiences as he grows up with 5-alpha reductase deficiency.

"It's so funny. I meet little old ladies who—when they find out what I do—say, 'Oh, we read about that in my book club. I totally understand what you're talking about.' And they do. That book made a difference." [7]

In 2007 Oprah Winfrey chose *Middlesex* for her book club. Because of that, millions of people have come to know more about intersex and the people it affects.

We are moving forward one step at a time.

EPILOGUE: UNTYING THE KNOT

When we divide the world into two groups, male and female, we tend to see all males as being similar and all females as being similar, and the two categories of "male" and "female" as being very different from each other. In real life, the characteristics of women and men tend to overlap. Unfortunately, however, gender polarization often creates an artificial gap between women and men.

> —from a publication of the National
> Honor Society in Psychology[1]

We still see a gap where none exists, a mirage that shimmers over the hot land of sex. And that mirage haunts us all, cleaves our lives in two. Real men are all man, and real women are all woman—Mars and Venus, Budweiser and Virginia Slims.

In truth, all of us fall somewhere in between our masculine and feminine ideals. Graphically the human race might look something like this:

Woman • Man

Each dot represents an individual. Of course, it would take a lot more dots to do this properly, since no two of us are identical, even with regard to sex. And, of course, the endpoints of this graph are hypotheticals, ideas, mental constructs, not real people. For some reason, we choose to call only the people who fall near the dead center of this chart intersex. But the center is just as essential as any other part of the continuum—without the middle, neither end is possible. And the middle really has no obvious boundaries.

In truth, we are all intersex, living somewhere in the infinite, but punctuated, stretch between MAN and WOMAN. But the idea of two opposite sexes is so ingrained in us that even seemingly educated people like U.S. presidents, senators, and members of Congress have argued that we should allow legally recognized marriages between only "a man and a woman." Never mind that it is impossible to define exactly what a man or a woman is.

What chromosomes would such people have? We've already seen that men, by many definitions, may have an X and a Y chromosome, two X chromosomes, two X and one Y chromosome, three X and one Y chromosome, and so on. By some criteria, there are also very real women with Y chromosomes in each of their cells, and others with Y chromosomes in only half of their cells. There is no specific combination of chromosomes or genes that unequivocally defines a "real" man or woman.

What genitalia should such people have? Many 46,XX persons have external genitalia that closely resemble those of men and internal genitalia that most closely resemble those of women. Even the combination of X and Y chromosomes with testes certainly does not define a man, since women with AIS have all of that and by other criteria are clearly female. Nor could we easily rely on the Phall-O-Meter. It seems obvious that no group, political or otherwise, could ever agree on exactly what constitutes a clitoris or a penis, especially if it has to be one or the other.

What sort of hormones should the fully male man have? Testosterone? During development and at birth, women with AIS have as much testosterone as any manly man, but by other standards they are not men.

Intersex people are not a few freakish, unfortunate outliers. They are instead the most complete demonstration of our humanity. Not one of the few I have been fortunate enough to come to know has suggested that he or she would rather be otherwise, although I am sure they haven't always felt that way. We, as a society, are very hard on people who don't fit with our preconceptions, especially our preconceptions about sex.

What intersex people have shown us is the truth about all of us. There are infinite chemical and cellular pathways to becoming human. Because of that, no two of us are now, or ever were, identical.

Sex isn't a switch we can easily flip between two poles. Between those two imaginary poles lies an infinite number of possibilities. Somewhere within that infinity is where you will find each of us. Intersex people have shown us that. We should be grateful. Because they are not bound by the traditional ropes of our traditions, they have shown us that we can untie the knots that bind us to our own preconceptions and begin to live freer lives.

ENDNOTES

Chapter 1: The Problem of Intersex

1. Tibbs, B. S. 1966. "An Approach to the Problem of Intersex: The Role of the General Pediatrician." *Pediatrics* 38 (3): 4305.

Chapter 2: A Brief History of Sex

1. Laqueur, T., *Making Sex* (Cambridge, MA: Harvard University Press, 1990).
2. Ibid.
3. Kemp, M. 1971. "Il Concetto dell' anima, in Leonardo's Early Skull Studies." *Journal of the Warburg and Courtald Institutes* 34: 115–134.
4. Ibid.
5. Laqueur, *Making Sex*.
6. Crick, F. *The Astonishing Hypothesis* (New York: Scribner, 1995).
7. Heuer, Richards J., Jr. *The Psychology of Intelligence* (Center for the Study of Intelligence, CIA, 1999). https://www.cia.gov/library/center-for-the-study-of-intelligence/csi-publications/books-and-monographs/psychology-of-intelligence-analysis/art5.html.
8. Laqueur, *Making Sex*.

9. Ibid.

10. Ibid.

11. Ibid.

12. Dickinson, R., *Human Sex Anatomy: A Topographical Hand Atlas*, 2nd ed. (London: London, Balliere, Tindal and Cox, 1949).

13. Johnson, V., Masters, W. H., and Lewis, K. C., *Manual of Contraceptive Practice* (Baltimore: Williams and Wilkins, 1964); Masters, W., Johnson, V. B. E., *Human Sexual Response* (Boston: Little Brown, 1966).

14. Riley, A., Lees, W., and Riley, E. J., "An Ultrasound Study of Human Coitus," in *Sex Matters*, ed. Bezemer, W., Cohen-Kettenis, P., Slob, K., and Van Son-Schoones, N. (Amsterdam: Elsevier, 1992).

15. Schultz, W. W., et al. 1999. "Magnetic Resonance Imaging of Male and Female Genitals During Coitus and Female Sexual Arousal." *BMJ* 319 (7225): 1596–1600.

16. White, M. March 2, 2006. "Two Thymuses Are Better than One." *Science Now* 3, http://sciencenow.sciencemag.org/cgi/content/full/2006/302/3 (requires registration).

17. Gage, F. H. 2000. "Mammalian Neural Stem Cells." *Science* 287 (5457): 1433–1438.

18. Bukovsky, A., et al. 2005. "Oogenesis in Adult Mammals, Including Humans: A Review." *Endocrine* 26 (3): 301–316; Bukovsky, A., Svetlikova, M., and Caudle, M. R. 2005. "Oogenesis in Cultures Derived from Adult Human Ovaries." *Reproductive Biology and Endocrinology* 3: 17.

19. Brisson, L., *Sexual Ambivalence: Androgyny and Hermaphroditism in Graeco-Roman Antiquity* (Berkeley: University of California Press, 2002).

20. Ibid.

21. Ellis, A. 1945. "The Sexual Psychology of Human Hermaphrodites." *Psychosomatic Medicine* 7: 108–125.

22. Money, J., Hampson, J. G., and Hampson, J. L. 1955. "An Examination of Some Basic Sexual Concepts: The Evidence of Human Hermaphroditism." *Bulletin of the Johns Hopkins Hospital* 97 (4): 301–319.

23. Money, J., Hampson, J. G., and Hampson, J. L. 1955. "Hermaphroditism: Recommendations Concerning Assignment of Sex, Change of Sex and Psychologic Management." *Bulletin of the Johns Hopkins Hospital* 97 (4): 284–300.

24. Meyer-Bahlburg, H. F. 1999. "Gender Assignment and Reassignment in 46,XY Pseudohermaphroditism and Related Conditions." *Journal of Clinical Endocrinology and Metabolism* 84 (10): 3455–3458.

25. Phoenix, C. H., et al. 1959. "Organizing Action of Prenatally Administered Testosterone Propionate on the Tissues Mediating Mating Behavior in the Female Guinea Pig." *Endocrinology* 65: 369–382.
26. Ehrhardt, A. A., Evers, K., and Money, J. 1968. "Influence of Androgen and Some Aspects of Sexually Dimorphic Behavior in Women with the Late-Treated Adrenogenital Syndrome." *Johns Hopkins Medical Journal* 123 (3): 115–122; Ehrhardt, A. A., Epstein, R., and Money, J. 1968. "Fetal Androgens and Female Gender Identity in the Early-Treated Adrenogenital Syndrome." *Johns Hopkins Medical Journal* 122 (3): 160–167.
27. Saez, J. M., et al. 1971. "Familial Male Pseudohermaphroditism with Gynecomastia Due to a Testicular 17-ketosteroid Reductase Defect. I. Studies in Vivo." *Journal of Clinical Endocrinology and Metabolism* 32 (5): 604–610; Saez, J. M., et al. 1972. "Further in Vivo Studies in Male Pseudohermaphroditism with Gynecomastia Due to a Testicular 17-ketosteroid Reductase Defect (Compared to a Case of Testicular Feminization)." *Journal of Clinical Endocrinology and Metabolism* 34 (3): 598–600; Goebelsmann, U., et al. 1973. "Male Pseudohermaphroditism Due to Testicular 17-hydroxysteroid Dehydrogenase deficiency." *Journal of Clinical Endocrinology and Metabolism* 36 (5): 867–879.
28. Imperato-McGinley, J., et al. 1974. "Steroid 5-Alpha Reductase Deficiency in Man: An Inherited Form of Male Pseudohermaphroditism." *Science* 186 (4170): 1213–1215; Imperato-McGinley, J., et al. 1979. "Male Pseudohermaphroditism Secondary to 5-Alpha Reductase Deficiency—A Model for the Role of Androgens in Both the Development of the Male Phenotype and the Evolution of a Male Gender Identity." *Journal of Steroid Biochemistry* 11 (1B): 637–645.

Chapter 3: Sex Versus Reproduction

1. Dawkins, R., *The Selfish Gene* (Oxford: Oxford University Press, 1976).
2. Morris, D., *The Naked Ape* (New York: Dell, 1967).
3. Wescott, R., in *Culture: Man's Adaptive Dimension*, ed. A. Montague (Oxford: Oxford University Press, 1968).
4. De Waal, F., *Bonobo: The Forgotten Ape* (Berkeley: University of California Press, 1997).
5. Savage-Rumbaugh, S., and Wilkerson, B. 1978. "Socio-Sexual Behavior in Pan paniscus and Pan troglodytes: A Comparative Study." *Journal of Human Evolution* 7: 327–344.

Chapter 4: Where Our Sexes Come From: The Abridged Version

1. MacLaughlin, D. T., and Donahoe, P. K. 2004. "Sex Determination and Differentiation." *New England Journal of Medicine* 350 (4): 367–378.
2. Grumbach, M. M. 2002. "The Neuroendocrinology of Human Puberty Revisited." *Hormone Research* 57 Suppl 2: 2–14.
3. Wilson, J., Foster, D. W., Kronenberg, M. D., and Larsen, R. P., *Williams Textbook of Endocrinology* (Philadelphia: W. B. Saunders, 1998).
4. Staub, N. L., and De Beer, M. 1997. "The Role of Androgens in Female Vertebrates." *General and Comparative Endocrinology* 108 (1): 1–24.

Chapter 5: Where Our Sexes Come From: The Rest of the Story

1. Grumbach, M., and Conte, F., "Disorders of Sexual Differentiation," in *Williams Textbook of Endocrinology*, ed. by Wilson, J., Foster, D. W., Kronenberg, M. D., and Larsen, R. P. (Philadelphia: W. B. Saunders, 1998).
2. Klinefelter, H. F., Jr., Reifenstein, E. C., Jr., and Albright, F. 1942. "Syndrome Characterized by Gynaecomastia, Aspermatogenesis Without A-Leydigism and Increased Excretion of Follicle-Stimulating Hormone." *Journal of Clinical Endocrinology* 2: 615–627.
3. Chen, H., "Klinefelter Syndrome," 2005. http://www.emedicine.com/PED/topic1252.htm. Accessed 2007.
4. Morton, C., Miron, P., "Cytogenetics in Reproduction," in *Reproductive Endocrinology*, ed. by Yen, S., Jaffe, R. B., and Barbieri, R. L. (Philadelphia: W. B. Saunders, 1999).
5. Jaffe, R., "Disorders of Sexual Development," in *Reproductive Endocrinology*, ed. by Yen, S., Jaffe, R. B., and Barbieri, R. L. (Philadelphia: W. B. Saunders, 1999).
6. Postellon, M., "Turner Syndrome," 2007. http://www.emedicine.com/ped/topic2330.htm. Accessed 2007.
7. Cotran, R., Kumar, V., and Collins, T., *Pathologic Basis of Disease* (Philadelphia: W. B. Saunders. 1999).
8. Jaffe, "Disorders of Sexual Development."
9. Grumbach, "Disorders of Sexual Differentiation."
10. McCarty, B. M., et al. 2006. "Medical and Psychosexual Outcomes in Women Affected by Complete Gonadal Dysgenesis." *Journal of Pediatric Endocrinology and Metabolism* 19 (7): 873–877.

11. Andersson, M., Page, D. C., and de la Chapelle, A. 1986. "Chromosome Y-Specific DNA Is Transferred to the Short Arm of X Chromosome in Human XX Males." *Science* 233 (4765): 786–788.

12. Titus-Ernstoff, L., et al. 2001. "Long-Term Cancer Risk in Women Given Diethylstilbestrol (DES) During Pregnancy." *British Journal of Cancer* 84 (1): 126–133.

13. Wilson, T., "Congenital Adrenal Hyperplasia," 2006. http://www.emedi-cine.com/ped/topic48.htm. Accessed 2007.

14. Glickman, S. E., et al. 2006. "Mammalian Sexual Differentiation: Lessons from the Spotted Hyena." *Trends in Endocrinology and Metabolism* 17 (9): 349–356.

15. Glickman, S. E., Short, R. V., and Renfree, M. B. 2005. "Sexual Differentiation in Three Unconventional Mammals: Spotted Hyenas, Elephants and Tammar Wallabies." *Hormones and Behavior* 48 (4): 403–417; Frank, L. 1986. "Social Organization of the Spotted Hyaena." *Animal Behavior* 35: 1510–1527.

16. Meadows, R. 1995. "Sex and the Spotted Hyena." *ZooGoer: Smithsonian National Zoological Park* 24: 1.

17. Glickman, "Sexual Differentiation in Three Unconventional Mammals."

18. Meadows, "Sex and the Spotted Hyena."

19. Lubahn, D. B., et al. 1989. "Structural Analysis of the Human and Rat Androgen Receptors and Expression in Male Reproductive Tract Tissues." *Annals of the New York Academy of Sciences* 564: 48–56.

20. Grumbach, "Disorders of Sexual Differentiation."

21. Adapted from Quigley, C. A., DeBellis, A., Marschke, K. B., El-Awady, M. K., Wilson, E. M., and French, F. S. 1995. "Androgen Receptor Defects: Historical, Clinical, and Molecular Perspectives." *Endocrine Review* 16 (3): 282. With permission.

22. Robertson, D. 1972. "Social Control of Sex Reversal in Coral-Reef Fish." *Science* 177: 1007–1009.

23. Ross, R., Losey, G. S., and Diamond, M. 1983. "Sex Change in a Coral Reef Fish: Dependence of Stimulation and Inhibition on Relative Size." *Science* 221: 574–575; Hamaguchi, Y., Sakai, Y., Takasu, F., and Shigesada, N. 2002. "Modeling Spawning Strategy for Sex Change Under Social Control in Haremic Angelfishes." *Behavioral Ecology* 13: 75–82; Shapiro, D. 1980. "Serial Female Sex Changes After Simultaneous Removal of Males from Social Groups of a Coral Reef Fish." *Science* 209: 1136–1137.

24. Adler, T., "Fishy Sex," 1995. http://www.sciencenews.org/pages/sn_edpik/
ls_4.htm. Accessed 2008.

25. Ibid.

Chapter 6: What We Do About the Ambiguous Child

1. Lee, P. A., et al. 2006. "Consensus Statement on Management of Intersex
Disorders. International Consensus Conference on Intersex." *Pediatrics*
118 (2): 488–500.

2. Huffman, J. W. 1976. "Office Gynecology: Some Facts About the Clitoris."
Postgraduate Medicine 60 (5): 245–247.

3. Money, J., Ehrhardt, A. A., *Man and Woman, Boy and Girl: The Differentiation
and Dimorphism of Gender Identity from Conception to Maturity* (Baltimore,
MD: Johns Hopkins Press, 1972); Money, J. 1998. "Case Consultation:
Ablatio Penis." *Medical Law*, 17 (1): 113–123.

4. Triea, K. 1998. "Power Orgasm and the Psychohormonal Unit." *Chrysalis*
5: 23–24.

5. Daaboul, J., and Frader, J. 2001. "Ethics and the Management of the
Patient with Intersex: A Middle Way." *Journal of Pediatric Endocrinology and
Metabolism* 14 (9): 1575–1583.

6. Lee, "Consensus Statement."

7. Frader, J., et al. 2004. "Health Care Professionals and Intersex Conditions." *Archives of Pediatrics and Adolescent Medicine* 158 (5): 426–428.

8. Weil, E. 2006. "What If It's (Sort of) a Boy and (Sort of) a Girl?" *New York
Times.*

9. Lee, "Consensus Statement."

10. Schober, J., "Feminization (Surgical Aspects)," in Stringer, M., Oldham,
K., and Mouriquand, P., eds., *Pediatric Surgery and Urology: Long-Term Outcomes* (New York: Cambridge University Press, 2006).

11. Ibid.

12. Ibid.

13. Rangecroft, L. 2003. "Surgical Management of Ambiguous Genitalia."
Archive of Disease in Childhood 88 (9): 799–801.

14. Creighton, S. M. 2004. "Adult Female Outcomes of Feminising Surgery
for Ambiguous Genitalia." *Pediatric Endocrinology Reviews* 2 (2): 199–202;

Creighton, S. M., Minto, C. L., and Steele, S. J. 2001. "Objective Cosmetic and Anatomical Outcomes at Adolescence of Feminising Surgery for Ambiguous Genitalia Done in Childhood." *Lancet* 358 (9276): 124–125.

15. Rink, R. C., Adams, M. C., and Misseri, R. 2005. "A New Classification for Genital Ambiguity and Urogenital Sinus Anomalies." *BJU International* 95 (4): 638–642.

16. Rink, R. C., et al. 2006. "Use of the Mobilized Sinus with Total Urogenital Mobilization." *Journal of Urology* 176 (5): 2205–2211.

17. Bettocchi, C., Ralph, D. J., and Pryor, J. P. 2005. "Pedicled Pubic Phalloplasty in Females with Gender Dysphoria." *BJU International* 95 (1): 120–124.

Chapter 7: Outcomes

1. Colapinto, John, *As Nature Made Him: The Boy Who Was Raised as a Girl* (New York: HarperCollins, 2000).

2. Ibid.; 1997. "Changing Sex Is Hard to Do." *Science* 275: 1475; Diamond, M., and Sigmundson, H. K. 1997. "Sex Reassignment at Birth: Long-Term Review and Clinical Implications." *Archives of Pediatrics and Adolescent Medicine* 151 (3): 298–304.

3. Colapinto, *As Nature Made Him*.

4. Zucker, K. J. 1999. "Intersexuality and Gender Identity Differentiation." *Annual Review of Sex Research* 10: 1–69.

5. "Changing Sex Is Hard"; Diamond, "Sex Reassignment"; Meyer-Bahlburg, H. F. 1999. "Gender Assignment and Reassignment in 46,XY Pseudohermaphroditism and Related Conditions." *Journal of Clinical Endocrinology and Metabolism* 84 (10): 3455–3458.

6. Colapinto, J. 2004. "Gender Gap." www.slate.com/id/2101678/.

7. Reiner, "Discordant Sexual Identity."

8. Meyer-Bahlburg, "Gender Identity."

9. Migeon, "Ambiguous Genitalia"; Migeon, "46, XY."

10. Hines, M., Ahmed, S. F., and Hughes, I. A. 2003. "Psychological Outcomes and Gender-Related Development in Complete Androgen Insensitivity Syndrome." *Archives of Sexual Behavior* 32 (2): 93–101.

11. Minto, C. L., et al. 2003. "The Effect of Clitoral Surgery on Sexual Outcome in Individuals Who Have Intersex Conditions with Ambiguous Genitalia: A Cross-Sectional Study." *Lancet* 361 (9365): 1252–1257.

12. Schober, J., "Feminization (Surgical Aspects)," in *Pediatric Surgery and Urology: Long-Term Outcomes,* ed. by Stringer, M., Oldham, K., and Mouriquand, P. (New York: Cambridge University Press, 2006).

13. Morgan, J. F., et al. 2005. "Long-Term Psychological Outcome for Women with Congenital Adrenal Hyperplasia: Cross-Sectional Survey." *BMI* 330 (7487): 340–341, discussion 341.

14. Cull, M. 2002. "Treatment of Intersex Needs Open Discussion." *BMI* 324 (7342): 919.

15. Creighton, S. M. 2004. "Long-Term Outcome of Feminization Surgery: The London Experience." *BJU International* 93 Suppl 3: 44–46.

16. Ibid.

17. Reiner, W. G., and Gearhart, J. P. 2006. "Anxiety Disorders in Children with Epispadias-Exstrophy." *Urology* 68 (1): 172–174; Diamond, "Sex, Gender, and Identity."

18. Diamond, M. 2004. "Sex, Gender, and Identity over the Years: A Changing Perspective." *Child and Adolescent Psychiatric Clinics of North America* 13 (3): 591–607, viii.

19. Diamond, "Sex, Gender, and Identity."

Chapter 8: Alternatives

1. Nanda, S., *Gender Diversity* (Long Grove, IL: Waveland Press, 2000).

2. Ibid.

3. Roughgarden, J., *Evolution's Rainbow* (Berkeley, CA: University of California Press, 2004); Nanda, *Gender Diversity.*

4. Jacobs, S., Wesley, T., and Lang S., eds. *Two-Spirit People* (Urbana, IL: University of Illinois Press, 1986).

5. Nanda, *Gender Diversity.*

6. Mahalingam, R. 2003. "Essentialism, Culture, and Beliefs About Gender Among the Aravanis of Tami Nadu, India." *Sex Roles* 49: 489–496.

7. Jacobs, *Two-Spirit People.*

8. Kuhnle, U., and Krahl, W. 2002. "The Impact of Culture on Sex Assignment and Gender Development in Intersex Patients." *Perspectives in Biology and Medicine* 45 (1): 85–103.

9. Ibid.

Chapter 9: Choices

1. Interview with author.
2. Ibid.
3. Ahmed, S. F., Morrison, S., and Hughes, I. A. 2004. "Intersex and Gender Assignment: The Third Way?" *Archives of Disease in Childhood* 89 (9): 847–850.
4. Ibid.
5. Zucker, K. J. 2002. "Intersexuality and Gender Identity Differentiation." *Journal of Pediatric and Adolescent Gynecology* 15 (1): 3–13; Zucker, K. J. 1999. "Intersexuality and Gender Identity Differentiation." *Annual Review of Sex Research* 10: 1–69.
6. Interview with author.
7. Ibid.

Epilogue

1. Matlin, M. 1999. "Bimbos and Rambos: The Cognitive Basis of Gender Stereotypes." *Eye on Psi Chi* 3: 13–14.

GLOSSARY

androgen insensitivity syndrome—complete and partial

A disorder of sex development that results from the lack of cell-surface receptors for androgens. AIS may arise as either complete or partial insensitivity. 46,XY infants with complete androgen insensitivity syndrome present as normal baby girls. Partial disorders may span the range between male and female.

androgens

Steroid hormones that are produced by the adrenal glands and that affect the development of male sex characteristics, including testosterone derived from progesterone.

androgyne, androgyny, androgynous

An organism having both male and female sexual characteristics and organs; at birth an unambiguous assignment of male or female is not possible. Androgyny is the union of two sexes in one individual. Androgynous is the adjective form.

175

androstenedione

A steroid hormone from which the body derives both male and female sex hormones.

clitorectomy

A surgical procedure in which the clitoris is removed.

clitoroplasty

Surgical alteration of the clitoris, often to reduce a large clitoris to a more standard size.

cloacal exstrophy

A birth defect that results in malformation of the pelvic region and associated internal organs. The cloaca is a common passageway for feces, urine, and reproduction found in some adult animals such as birds. But during human development, the cloaca forms, then usually disappears later as the individual parts separate to form a rectum, a bladder, and genitalia. In exstrophy of the cloaca, the cloaca never fully separates into the individual parts, so the lower region of the abdomen doesn't fuse, and what would normally become the bladder, etc., is open to the outside. Cloacal exstrophy affects boys and girls but is most severe in boys, because boys with cloacal exstrophy have cryptorchidism (failure of the testes to descend into the scrotum) and very severe epispadias (a malformation of the penis in which the urethral opening is above the phallus).

congenital adrenal hyperplasia

This disorder is also known as adrenogenital syndrome or 21-hydroxylase deficiency. Congenital adrenal hyperplasia covers a range of inherited disorders of the adrenal gland. It can affect either boys or girls. People with congenital adrenal hyperplasia partially or completely lack an enzyme that aids in production of cortisol and aldosterone. Under these conditions the adrenal glands produce much higher levels of androgens, which causes male characteristics to appear early (or inappropriately).

dihydrotestosterone

A steroid hormone derived from testosterone using the enzyme 5-alpha reductase. Along with testosterone, this hormone is essential for normal masculinization of the external genitalia of male fetuses.

disorders of sex development

A catchall phrase to describe any alterations in the normal pattern of development of internal and external genitalia in animals, mostly humans.

estrogen

A steroid hormone that plays a role during primary (present at birth) and secondary (arises at puberty) sex development in females.

Fallopian tube

Either of a pair of ducts that deliver ova from the ovaries to the uterus in the female reproductive system of humans and most other mammals. Also the site of fertilization.

gender identity

A person's own sense of being male or female. Gender identity may be unrelated to the person's external genitalia or karyotype.

gender reassignment

In children, this refers to surgical and hormonal procedures usually undertaken to give a 46,XY male child the sexual characteristics of a female. Usually only attempted when, for one reason or another (such as cloacal exstrophy), traditional development as a male is not an option.

genitoplasty

Surgical alteration of the external genitalia. Usually done to make a person more closely resemble an average male or female.

gonadal dysgenesis—pure and mixed, complete and partial

Abnormal development of one or more gonads. Complete failure of gonadal development is known as pure or complete gonadal dysgenesis. Partial development of the gonads is known as partial or mixed gonadal dysgenesis.

gonads

Organs inside plants and animals that produce gametes—either ovaries or testes producing ova or sperm, respectively.

hermaphrodite; true hermaphrodite; pseudohermaphrodite

An animal or plant with reproductive tissues of both sexes. In humans, this is a person with one testis and one ovary or one or more ovotestes. Traditionally, the term "true hermaphrodite" has been used to refer to people with gonadal tissue of both sexes—either one testis and one ovary, or one or two ovotestes. "Pseudohermaphrodite" has been used to refer to people whose external genitalia are not consistent with their chromosomal makeup. Most pediatric endocrinologists consider this term archaic and potentially offensive when use to refer to a human being.

hypospadias

Most commonly described in males, this disorder of development results in the urethral opening forming somewhere between the glans and the base of the penis instead at the tip. In females, it may result in the urethra opening into the vagina.

intersex

Human beings who cannot be easily classified as either male or female.

karyotype

The complement of chromosomes found in an organism. For humans, the usual karyotype is 46,XY or 46,XX—human beings usually have a total of 46 chromosomes including either one X and one Y sex chromosome (male) or two X sex chromosomes (female).

Klinefelter syndrome

A genetic variation in males that usually results in a karyotype of 47,XXY. Except for sterility, this change results in few or no physical or mental alterations. As a result, most men with this condition are not diagnosed unless they seek fertility counseling as adults. Klinefelter syndrome also includes other genetic variations that result from multiple X chromosomes in association with a Y chromosome, such as: 48,XXXY, 49,XXXXY, etc. Increasing numbers of X chromosomes are accompanied by increased physical abnormalities including mental retardation.

meiosis

A process of cell division in sexually reproducing organisms that reduces the total number of chromosomes from the normal diploid state to half that number (haploid) and results in the production of sperm and eggs.

mitosis

The most common form of cell division. It results in two identical daughter cells with an equal, diploid number of chromosomes.

oogenesis

The biological process that results in the production of ova (which begin their lives as oocytes).

orchidectomy

Surgical removal of one or both testes. Sometimes also called an orchiectomy.

ovotestes

Gonads that include tissues characteristics of both ova and testes.

phalloplasty

Surgical construction of or alteration of the penis. Usually done to make a person more closely resemble an average male.

pseudohermaphrodite

A catchall word to describe disorders of sexual development that do not result in true hermaphrodites. It includes individuals with internal genitalia of one sex and external genitalia of the other, or individuals in which the karyotype does not correspond to the appearance of the external genitalia.

sex chromosomes

In most mammals, the sex chromosomes are the chromosomes that determine the sex of the individual. In humans these are the X and Y chromosomes: 46, XX usually being female and 46,XY usually being male.

sex-determining region of the Y chromosome

The region on the Y chromosome that contains the genes necessary to make a developing fetus into a male. This region is also known as SRY (sex region Y).

spermatogenesis

The biological process that results in the production of sperm (which begin life as spermatocytes).

testosterone

A steroid hormone produced primarily in the testes (but also to a lesser extent from adrenal androgens) that is responsible of the production and maintenance of male sexual characteristics, including the external and internal genitalia along with the production of sperm.

Turner syndrome

A chromosomal variation in females that is characterized by the absence of all or part of a second sex chromosome in some or all cells. The physical features of women with Turner syndrome often include webbing of the neck, short stature, shortened fourth and fifth fingers, broad chest, cardiovascular abnormalities, and gonadal dysgenesis (streak ovaries). Women with Turner syndrome are usually infertile due to failure of ovarian development.

urogenital ridge

A pair of fetal structures that appear early during development. A urogenital ridge contains material from which the gonads, the external and internal genitalia, the kidneys, the adrenal glands, and the urogenital system will develop.

urogenital sinus

Also known as the persistent cloaca, the urogenital sinus is a part of the human body usually present only during the development of the urinary and reproductive organs. It is the lower part of the cloaca, formed after the cloaca separates from the rectum. It eventually becomes, among other things, the bladder. A urogenital sinus anomaly is also a rare birth defect in women where the urethra and vagina both open into a common channel.

vaginoplasty

A surgical procedure that is performed to reshape, refine, reconstruct, or construct a vagina in human beings. In children, the most common goal of vaginoplasty is to construct or refine a vagina to a point where it will function in normal sexual intercourse.

Resources

Web Sites

Accord Alliance
www.accordalliance.org

A nonprofit organization established by a national group of health care and advocacy professionals to promote comprehensive, integrated approaches to care that enhance the health and well-being of people and families affected by disorders of sex development. The group fosters collaboration among all involved and works to ensure that new ideas about best care are disseminated and implemented across the United States. Accord Alliance does not provide direct support services, but the site does provide links to resources for families and patients including books, organizations, and a list of support groups.

Androgen Insensitivity Syndrome Support Group (AISSG)
www.aissg.org

The following is a partial list of topics at this UK site: What is AIS?, Complete AIS, Partial AIS, Related Conditions, Group Meetings, Raising Awareness, Personal Stories, Obtaining/Facing Diagnosis, Orchidectomy (Gonadectomy), HRT/Osteoporosis, Vaginal Hypoplasia, Genital Plastic Surgery, Recommended Clinicians, Research Studies, and Fertility Advances. Much of the information is available in many languages besides English.

Bodies Like Ours

www.bodieslikeours.org

This site offers several forums for discussion of issues related to disorders of sexual development including, but not limited to, androgen insensitivity syndrome, congenital adrenal hyperplasia, Klinefelter syndrome, Turner syndrome, and several others as well as general forums for discussion of any issues related to DSDs.

DSD Guidelines

www.dsdguidelines.org

This project of the Consortium on Disorders of Sex Development offers two monographs related to DSDs: *Clinical Guidelines for the Management of Disorders of Sex Development in Childhood* and *Handbook for Parents*. These monographs have been produced by a consortium consisting mainly of: (1) clinical specialists with experience helping patients with DSDs; (2) adults with DSDs; and (3) family members (especially parents) of children with DSDs. This site also offers several links to support groups and portraits of individuals with DSDs.

GendersInx.org

www.gendersinx.org

A support forum that encourages all involved, including parents of children with Turner syndrome, Klinefelter syndrome, androgen insensitivity syndrome, congenital adrenal hyperplasia, and 5-alpha reductase. The organization is run by those who themselves have the conditions and believes in positive reinforcement for better education.

Intersex Society of North America

http://isna.org

A comprehensive site that contains a wealth of information about the character of intersex, several personal stories, their recommendations for assessing and dealing with health issues related to disorders of sexual development, tips for parents, a teaching kit to accompany undergraduate courses that discuss intersex, and links to a comprehensive list of support groups.

Organization Intersex International—USA (OII-USA)

www.intersexualite.org/usa.html

The U.S. affiliate of the Organisation Intersex International, a nonprofit organization incorporated in Quebec, Canada. OII-USA devotes itself to preventing genital mutilation, medicalization, and normalization without consent and offers another face to intersex lives and experience by highlighting the richness and diversity of intersex identities and culture. According to this site, OII is the largest intersex organization in the world.

A partial list of their offerings includes: Official Positions, FAQ on Intersex, FAQ About OII, Brochures, Medical Information, Online Support Group, Condition Specific Support Group, Fact Sheet on DSD Activism, Why Is OII Not Using the Term "DSD," and Intersexuality in the Family.

ParentsInX.Org

www.parentsinx.org

An online community connecting families living with congenital gender variations. This site offers several different forums for discussion of issues related to different genetic conditions including Klinefelter syndrome and Turner syndrome, and offers a handbook for parents written by parents.

Books

Dreger, Alice Domurat. *Hermaphrodites and the Medical Invention of Sex* (Cambridge, MA: Harvard University Press, 2000).

Dreger, Alice Domurat, ed. *Intersex in the Age of Ethics* (Hagerstown, MD: University Publishing Group, 2000).

Laqueur, Thomas. *Making Sex: Body and Gender from the Greeks to Freud* (Cambridge, MA: Harvard University Press, 1990).

Preves, Sharon E. *Intersex and Identity: The Contested Self* (Piscataway, NJ: Rutgers University Press, 2003).

People

Nicky Phillips

Contact Person for Androgen Insensitivity Syndrome: Canada
A contributor to this book and a woman with androgen insensitivity syndrome, and a resource for information about AIS and for individuals seeking information or help. Of all the people I interviewed, only Nicky offered me her address and phone number as a resource for helping others. I am most grateful to her for that.

#17, 3031 Williams Road, Richmond, BC, Canada V7E 1H9
Telephone: (604) 274-9630

INDEX